high tech BABIES

high tech BABIES

an owner's manual

How to encourage your child's early interest in science and math

Emily Needham Williams

PRESSWORKS

Copyright © 1986 Pressworks
Dallas, Texas

Library of Congress Cataloging-in-Publication Data

Williams, Emily Needham, 1952–
  High tech babies.

  Bibliography: p.
  Includes index.
    1. Children's questions and answers. 2. Science—
Study and teaching. 3. Mathematics—Study and teaching.
4. Domestic education. 5. Child development.
I. Title
HQ784.Q4W55   1986   649'.68   85-28101
ISBN 0-939722-23-2

Design by
The Smitherman Corporation

To the people who led me to a love of science: Jim and Beth Needham, Mr. A, Charlie, Betsy, Anna, and Mary.

Contents

Preface	ix
I. So Many Questions	1
II. The Developing Mind	13
1. Infants and Toddlers: Learning through Action	13

    Newborn / Four months / Seven months / Ten months / Accident proofing / Twelve months / Eighteen months

2. Preschoolers: Asking Questions	27

    Two years / Three years / Four years

3. Early Schoolers: Forming Ideas	37

    Five years / Six years / Seven years

III. Areas of Interest	45
1. Human Biology	45

Skin and Hair / Bones / Muscles / Blood / Breathing / Noses / Digestion / Teeth / Nerves / Reproduction / Death / Creation

2. Zoology	71

Body coverings / Scales / Feathers / Fur / Camouflage / Body parts / Animal behavior / Learned behavior / Ecology / Food chains / Decay / Communities / Parasites / Helpful relationships / Dangerous animals / Temporary guests / Animals and seasons

3. Botany	97

Photosynthesis / Leafy structures / Fruits, vegetables, and nuts / Roots / Seeds / Flowers / Dangerous plants

4. Microbiology	109

Fungus or mold / Bacteria / Viruses / Algae / Disease-producing organisms / Fever

5. Chemistry	115

Nutrition / Water supply and waste disposal / Solids, liquids, and gases / Chemical change / Substance misuse / Smoking

6. Physics	126

Sounds / Ears / Light / Color / Eyes / Weather / Wind / Clouds / Lightning and Thunder / Heat / Electricity / Simple Machines / Space

Inadvertent Lessons: Reading to your Child	157
Bibliography	161
Index	164

Preface

This is a one-of-a-kind book for parents of children from birth to eight years of age. It is designed to assist parents in maintaining and expanding a child's innate intellectual curiosity by matching the answers and responses to a child's questions and experimentations with the appropriate level of information for each stage in the child's early development.

For the past seven years, I have listened to the questions that very young children ask. I have watched as they investigate sound, weather, insects, and rocks, and wondered if all of this scientific investigation will have any effect on their appreciation of science later on in their lives. Mostly, I have questioned the part that parents play in their children's earliest explorations. I have tried to help friends when their children asked "Which end of the worm is the head" and other questions the parent wasn't confident in answering. I have watched as parents admonished their children to "Stay away from spiders—they bite!" I have listened to them answer a child's question with "You wouldn't understand," and I have wondered if we are promoting an alarming rise in scientific illiteracy by virtue of our own words and actions.

In 1982, the National Science Teachers Association published a position statement on Science Education for the 1980s. The statement stressed that the crisis in science education demanded immediate attention in order to avoid raising a generation of "scientifically illiterate" Americans. It proposed that science become an integral part of the elementary school curriculum. The report emphasized that elementary schools need science programs that promote an understanding of the world in which we live, as well as engender interest and appreciation in it. It was deemed necessary that children be given an opportunity to explore and investigate their world by coming in contact with it—the "hands-on" approach.

But the children *most* likely to explore and investigate their world are little children—those under the age of eight. No human age is more dedicated to constant exploration than children from ages one to six years. These children are the embodiment of curiosity; they want to know what makes things move, what sound is made when two objects collide, how things feel to the touch and why.

During the first seven years, it is parents who influence their children's exploration to the greatest degree. As Burton White observed in *The First Three Years of Life*,

> We came to believe that the informal education that families provide for their children makes more of an impact on a child's total educational development than the formal educational system.

The National Commission on Excellence in Education spoke directly to parents in its report to the nation called, *A Nation at Risk*:

> As surely as you are your child's first and most influential teacher, your child's ideas about education and its significance begin with you. You must be a "living" example of what you expect your children to honor and emulate.

Robert Yager of the Science Education Center at the University of Iowa cites numerous studies concerning the crisis in science education and concludes that the major crisis in science education is a failure to bring about scientific and technological literacy among all citizens. He also stressed that *parents and peers*, and possibly television, rank more importantly than schools in promoting knowledge and interest in science.

It is the parents who promote their young children's interest in science. In her book *Growing up with Science*, Marianne Besser interviewed the parents of the Science Talent Search winners, high school students who, in the late 1950s, showed extraordinary talent in the areas of science and mathematics. When the parents were asked how they felt they had contributed to the development of their child's talent during his or her early years, these parents consistently pointed to the same thing—a tendency to reward their children's earliest questions with interest and answers.

Their children were encouraged to investigate and explore interesting objects and events, to whatever extent they wanted, within the realms of safety. From their children's earliest years, these parents had placed a high premium on "finding out why" and extending the investigations to the farthest limits.

All parents appreciate their children's curiosity. They are also aware that creative thinking, problem-solving, and self-motivated investigation are tools necessary for success in modern society. But often parents fail to capitalize on their children's earliest attempts at understanding their environment by the manner in which the first questions are answered. Parents may feel inadequate in their own ability to explain things about the natural world and the physical principles which govern it. They may find it difficult to match the complexity of an answer with the ability of a child to understand. They may fail to recognize the very critical importance of rewarding a youngster's early attempts to question and understand the world of nature.

It is difficult to discern all the reasons it has become difficult for parents to answer their children's questions about basic scientific principles. Although their early science education is partly responsible, many sociological factors undoubtedly contribute to the problem. Standard of living increases often dictate two wage-earner families, leaving less time for parents to interact with their children. Entertainment tastes have changed in favor of television and sports and away from early and extended exposure to nature like long walks and fishing. The value placed on intellectual vitality has diminished, resulting in fewer dinner-table discussions about social and scientific dynamics. The paradox is that as society has become highly technological, food production and power sources have become physically remote and completely separated from the average citizen's daily life experience. Machines that require no understanding on the part of the operator cook the food, wash the dishes, allow communication with far-off loved ones, move people from place-to-place and entertain them. No longer do average families have daily chores involving soap making, energy generation, and food production. It is no longer necessary to study and understand weather cycles, or to log positions of the sun and stars for survival.

But while technological advancements provide changes in people's lives, the principles that govern the natural world remain the same. The same phenomena peak children's curiosity; the same questions appear as appeared before electricity, before telephones, before computers. The moon still seems to follow them at night; the leaves still turn brown and fall off the trees in the fall; the grass still feels cold and wet in the morning; the embers of a fire still glow brilliantly orange after the flame has disappeared. And now there are new questions—about the television, about electrical outlets, about airplanes, about medicine and "shots," about nuclear energy and warfare.

Before technology brought industrialization to American society, most people lived on the farm. In these families, children were viewed as small adults. There was work to be done in order to carry on an existence. Not only was the child's help needed, his future welfare depended on his ability to use the tools and information given to him by his family. School wasn't really necessary—what was necessary was producing food to eat and maintaining a shelter to live in. The future was essentially ordained by the past. Work was done a certain way because that was the way it had always been done. Children had a basic knowledge of, if not an understanding of, natural phenomena, and questions that arose were answered by attributing cycles and events to acts of God. Even though these children lacked extensive formal education and rarely envisioned themselves as anything separate from their own community and heritage, they nevertheless experienced nature firsthand. Their science education remained purely in the realm of experience.

When industrialization began to change the makeup of American society, families became less rural and more urban, and the child's position came to be defined quite differently. Parents were encouraged to promote expressions of competitiveness, of risk-taking, and of high achievement. It was important for every person to work hard and to study hard to prepare to gain, later on, his piece of the "American Dream." This was a forward-thinking, optimistic, future-oriented society in which young people understood that the promise of the future was one's own responsibility.

Education, during this time, was given the task of conveying the basic skills of reading, writing, and arithmetic, which were thought to be the tools needed to make future accomplishments. Science education was directed toward finding better ways of doing things, toward developing a "better way of life," toward alleviating some of the world's critical problems. Education was undeniably future-oriented, bent on training children to attain success and to become good citizens.

Students during the industrial age were benefiting from formal, specific training in the sciences. As children do, they were asking the questions, "Why does this happen?" and "How does this happen?" They were learning effective scientific techniques, and they were still not totally removed from the "hands-on" experience of their farming heritage. At the very least, Grandma and Grandpa still lived on the farm and needed their help during harvest season or enjoyed their company during summer vacation. Children picked the crops, carried the water, understood which clouds carried needed rain, and tinkered with the tractor.

The orientation toward the future that students had during the industrial age made them good scientists. It was easy for them to invest time and effort into both study and experimentation. They already understood that effort in the present would bring about payoffs in the future. They were willing to investigate the "before" and then document the "after" of the experiments. This method of investigation is essential to scientific inquiry.

Today's parents are the first modern generation to experience formal science education without considerable natural experience. When asked if they have relatives on the farm, most parents have to reach back at least two generations to find them. Socio-economic change has produced a society far removed from its farming background, and the ease with which people relate scientific principles to their own sphere has suffered from that change. In an urban society, children receive most of their understanding of the rules governing the natural and physical worlds from science classes in a school room.

The lack of firsthand experience with basic science principles is exacerbated by the fact that modern parents have machines that do most of their work, eliminating the necessity of invention. Understanding the principle of a pulley is unnecessary to operate an automatic garage door opener.

To ensure that the newest generation of Americans does not suffer from a lack of scientific literacy, parents need to bring science back to everyday living. In the Mid-1800s, a British scientist, Thomas Huxley, attempted to bring science to the people. Until that time, most people had tried to understand the world around them by applying old opinions and superstitions. Huxley held that science was nothing more than trained and organized thought and suggested the substitution of science's well-tested facts for the prevailing incorrect explanations. Huxley's recommendation can be applied to parents and children. When children earnestly seek the input of their parents in an attempt to understand an object or situation for which they have no explanation, parents should give them "science's well-tested facts." The information must be provided when it is asked for and in a form that the child will understand. Despite the fact that the body of scientific knowledge is so large that it threatens to leave the average citizen behind, much of science remains in the realm of common sense as building blocks to the complicated high tech future. And, after all, it is within the realm of common sense that parents communicate with their children.

It's easy to say that parents should begin to answer their children's questions with science's facts. It's much harder to tell them how to prepare to do so. After seven years of searching, I have not found a book that could prepare parents for their important role as facilitators of their children's early science experience. There were books of questions and answers, but no one can possibly anticipate all the questions that very small children ask. And even if you could, there would be no time to look up the answer. An encyclopedia of science facts wasn't enough. For even a command of scientific knowledge cannot prepare a parent for dealing with "Why is the moon following me?" Books of specific science activities designed to be performed by the child and parent failed to provide any general information pertaining to the child's everyday experiences.

Parents, I decided, needed a book that would encourage them to listen to their child and answer her questions, as well as promote her innate sense of curiosity and wonder. It would need to include a description of a child's way of thinking and understanding during each of the first seven years, how that thinking changes, and outward signs that signal its transposition. Parents would need to predict the typical interests of a child during each of these first seven years and understand what motivates her questions. Most of all, the book would need to provide basic scientific information in each of eight areas of interest to children under eight, along with sample questions and answers that were appropriately educational and consistent with the child's ability to understand.

I decided to write that book. Here is a manual for parents who want to gently prepare their young children to grab hold and take every advantage of the high tech world into which they were born.

I

So Many Questions

For your children, science is just plain fun. It's creative, like art, and allows children a real outlet for their creativity. It is playful, because it requires manipulating, mixing, building, and exploring. It's also organized, and children show a real tendency to appreciate organization. All children are natural scientists in the beginning, in both senses of the word "natural." They like to "make things fit." Science is very much like a puzzle—the whole picture doesn't really become clear until the individual pieces all fit together, and children always like puzzles.

Watching children play outside provides numerous examples of what captures their attention. They pick up objects that seem to have no value, like a dried, mashed frog scraped off the driveway or a discarded piece of cable wire. They throw rocks onto the pavement just to see them break—they are so much shinier inside! They stop and listen to interesting sounds—birds singing, squirrels chattering, trains whistling. They feel things, like how the sand is warm on the sunny side of the sandbox and cold on the shady side. They watch movements such as squirrels leaping from branch to branch or airplanes leaving condensation trails.

Science-related activities are natural extensions of children's intense creativity, and they often require input from adults. However, for many parents of small children, all the fun has gone out of science. It is perceived as being something cold, technical, difficult to understand—something having very little to do with everyday living. By defining science as all things "high tech," many people have forgotten the "low tech" aspects of everyday science. It is in the realm of "low tech" that children first show an interest in science, and it is the science perceived by children that has the greatest meaning in the daily lives of families. It is the exploring, the building, the cooking, and the problem-solving that generates curiosity in children.

The first questions that are asked of parents come from toddlers; their questions are generally limited to the "What's that?" category and are often asked as much in an effort to practice vocabulary as to gain information. Parents have very little problem with these kinds of questions. Eighteen to twenty-four-month old children require answers that are short and easy for the children to remember—moth, wheel, rain, or moon. Parents are delighted that their children are talking; it may not have occurred to them yet that their children are also showing an intense curiosity about science.

Gradually though, the "What's that?" questions of the toddler evolve into the demanding "Why?" questions of pre-schoolers and early schoolers. (One study found that three-year-olds ask more than three hundred questions a day.) In answering, parents are able to supply information and, at the same time, alleviate fear, extend attention, grant assurance, or separate reality from fantasy. Parents also have a powerful ability to encourage their children's interest in science, both natural and physical.

The problem is, many parents fail to perceive the hidden complexity in a child's simple questions. The reasons for that failure are many. They are certainly not indifference toward the intellectual development of the child, for most parents want their children to learn as much as possible. They may include a time-intense lifestyle that allows little time for listening to children's questions, and they may include a lack of interest in science on the part of the parents. But, without a doubt, the most important reasons why parents fail to answer little children's questions are that they are afraid they do not know the "right" answer or that they do not understand the developmental stage of their child and the child's real intellectual need for an explanation. Parents either feel that their own knowledge and understanding is lacking, leaving them without an answer, or parents feel they have an answer, but are unsure of the correct way to pass that information on to the child.

Whatever the reason in not answering the child to the fullest, the parent has missed an opportunity to encourage a deepening of his or her child's curiosity into the scientific world. Children ask more questions when they believe that they will receive answers. Conversely, if their questions are repeatedly shrugged off or left unanswered, eventually, fewer questions will be asked. Indifference toward curious behavior results in apathy in the child and the development of an attitude that simply accepts things the way they are.

The most important aspect of answering a child's question is to give the question importance—to answer it honestly and, when the answer doesn't satisfy, to seek out additional sources of information (identification books, science dictionaries, neighbors, or relatives versed in that area of interest). Then the child will know that her parents approve of, and encourage, a questioning attitude. The answer itself is not even as important as the attitude the parent expresses when the question is asked. Parents need to worry less about what constitutes a "right answer" and be more concerned with promoting their child's curiosity. Curious children have been shown to perform better in school, to be more creative, and to be quicker at problem-solving. There is no doubt that questions serve an important function in the development of the child, and the science-related questions of very young children are forming the foundation for a skill that will later be crucial to the education of the child—the understanding of cause and effect.

The first thing that parents must do in order to communicate with their young children about science is to try to see the world the way a child sees it (more on this in Chapter Two). Children see a "low tech" world, not a "high tech" one. The world children see today is much the same as the world children saw a century ago, or even a thousand years ago. They are fascinated by rainbows, insects, rocks, clouds—natural objects that have captured the attention of children for all of time. In order to take their children's view of the world, today's parents have to remove themselves (at least temporarily) from the "high tech" aspects of daily living and imagine what their children see. Often, in doing this, parents recapture their own interest in science and nature and, with new enthusiasm and confidence, can recall from their own experience enough information to answer their children's questions.

For one thing, children are much closer to the ground than are adults. By physically moving into the child's space, a parent can begin to appreciate their child's view of the world. When a child is actively observing an object or event, her parent can crouch down beside her and join in on the investigation. This helps the parent to understand the motivation behind the question.

The second point that parents must remember is that the reaction of an adult to a child's investigation can have a profound impact on the development of the child's attitude toward science in general. Children whose parents display an attitude which is unprejudiced and open-minded toward investigation, as well as free from superstition, are the most likely to develop those same attitudes. All of these attitudes can be promoted in children, even before the age of one year, if the parents understand the development of learning by imitation. It is the ability to imitate that influences a child's interaction with objects in nature. If a parent consistently expresses fear or distaste when approached by a child holding a harmless insect, the child, in imitation, may begin to react in the same manner. On the other hand, if the parent confronts the insect in a respectful, but inquisitive manner, the child will do the same. Even such attitudes as the tendency to challenge sources of information and the hesitancy to jump to conclusions are considered to be part of the scientific attitude and, therefore, should be demonstrated to very young children. Parents may need to investigate their own feelings about some of the tough questions involving such topics as sexuality, death, and fear of injury, since the attitude expressed by the parent contributes to the understanding gained by the child.

The third point that will help parents in their attempt to recognize and promote their child's early interest in science is to understand how their children think as well as to be able to predict the misconceptions that their children already hold. This information is expanded in Chapter Two.

The final aid to parents is a brush up on their own general knowledge of science and nature. As children of the fifties and sixties, today's parents, unfortunately, received a science education that was most likely a "read the chapter and answer the questions at the end" science education. Regardless of how well-intentioned that kind of science learning was, it failed to provide enough practical information about everyday science. Bits and pieces of information learned by reading and memorization are easily forgotten and provide little understanding of the "Big Picture." In other words, a large number of today's parents were not likely to have spent a great deal of time observing the pieces of the science puzzle and making them fit. That kind of knowledge is born of firsthand observation and manipulation. Fortunately, science education is becoming more experiential; children are now learning by doing and seeing, rather than merely by reading and writing. But modern parents are not likely to benefit from this change in method. In order to refresh their memories of early-learned facts and principles, they may wish to read the final chapter and re-learn a few of the basic principles governing life science and physical science.

Once parents are free from worrying about knowing "the right answer," they begin to respond to more of their children's questions. Some of the answers parents give may be inappropriately complex, but these answers generally do nothing harmful to the child (they do leave a feeling in the child that her question was important enough to deserve an answer). Sometimes, given a complex answer, children will even alter it to fit their own level of understanding. This can lead to inaccurate, and many times, humorous explanations. (A friend once answered her young daughter's question about the way that her baby brother or sister began growing by comparing the unborn baby to a growing seed. Her daughter proudly related this information to Daddy later that evening, informing him that Mommies eat seeds when they want to have a baby.)

Inevitably, there will be questions asked whose answers are far beyond the child's ability to understand. In these situations, it is perfectly correct to tell the child that the answer is too difficult or, better yet, to give the child some piece of understandable information about the object she finds interesting. Her questioning can then be diverted to something that will make more sense. In this way, the parent is moving with the child in the direction of her interest.

There will be times when parents need to answer "I don't know." No one can be expected to know everything. It's important for the children to see that their parents do not know everything but will make an effort to look up an answer. It teaches the children that there is a great deal to be known— so much, that even mothers and fathers haven't been able to learn it all. It also teaches them that life is a constant search for answers. By watching their parents search for understanding, they will come to see life as a learning experience (even for "grown-ups").

Finally, young children must feel safe to pursue the answer to a question, free from fear of failure or evaluation. Parents often find it difficult to accept their children's unsophisticated theories about how things work—theories the children are often unwilling to reconsider. Children are sometimes afraid to consider the possibility that a phenomenon that they were certain they understood actually takes place for a different reason. It is that fear that illicits the "No, you're wrong" response in children. But parents who provide a "safe" atmosphere for their curious children are always willing to accept reactions that are anything but standard; these parents are providing an opportunity for their children to lay the foundation for the reasoning process.

Asking questions of children is an effective means of stimulating curiosity. Despite the fact that children ask many questions, they are not automatically successful at asking "good" questions, the kind that lead to an investigation in an attempt to find an answer. Asking these kinds of questions requires intuition, a quick knowledge of or insight into an object, a truth, or a principle. Intuition requires cultivation; in order to question an object or a principle, something must be known about it. For that reason, it is often necessary for parents to ask their children questions about an object or event that the children find interesting. It may be that the children have so little information about the interesting object that they cannot even ask a question about it, or their questions can go no further than "What is that?" By asking their children questions which challenge already held notions, parents can lead children into active learning situations that provide much more information than simply a name.

Children are very good at asking "What is this?" But, given an answer, they do not often continue to investigate further through conversation, although they may be observed spending more time manipulating the object and practicing new skills on it. All of the experiences involving touching, tasting, lifting, throwing, bending, and balancing objects are building up a library of mental images that will be used to facilitate abstract thinking later on. But children also enjoy an extended conversation with parent or teacher concerning their curious objects. Through this conversation, children are rewarded for their curiosity with positive attention, making them much more likely to ask questions in the future. Also, their powers of observation are sharpened by leading them into a closer inspection of the object or event they found interesting. It also helps children become comfortable with words and learn their definitions through usage.

Children need to learn the vocabulary pertaining to a particular concept. Parents need not severely limit their vocabulary when they speak to their children for fear that the words will be too hard for children to understand. The experience comes first, the vocabulary follows. Vocabulary gives a mental picture of particular objects, events, or principles. It helps recall, and it simplifies complex things in order to tie them together. Vocabulary is critical but should never be given importance over understanding. In the case of giving vocabulary to young children, it is wise to avoid over-generalizing or using "baby talk." Instead, the child can be encouraged to learn the correct word or phrase to describe an object or event. For example, it is quite obvious to a child as young as two years of age that beetles, butterflies, and spiders are different animals. At eighteen months, the proper vocabulary for this entire group of animals might be "bug," but by two and one-half years, they should no longer be described as being bugs. Beetles are insects; spiders are spiders. Even butterflies and moths are generally easy to distinguish, as are worms and caterpillars. All that parents need to remember is that children are excellent observers, able to notice even slight differences in body shape, movement, or color. The child's powers of observation are strengthened through the use of correct vocabulary. Also, comparisons are more easily made when there are words to describe the differences.

Good questions to ask children motivate them to question further. These are the "why, how, and where" questions. ("Why did that lizard change colors? How do you think he did it? Where did he go when you tried to catch him?) But questions which cloud or distort the issue confuse children and leave them with negative feelings about the conversation. Therefore, it is best to avoid questions that contain difficult or unfamiliar words (Where did all of this desiccated material originate from?), questions which are vague and leave the child unsure of what is being asked (How do you feel about clouds?), and questions, that in the mind of the parent, have only one right answer (What is this water sticking to the grass called?). Remember that young children do not need instruction where they have no interest.

Children want to be like their parents, and when their parents question what they see and try hard to understand it, their children will do the same. Children are full of information, they just need someone to question them in order for them to draw on that information and then apply it to the situation at hand. Eventually, children begin to do this on their own, and make a giant step toward understanding the scientific method of investigation. However, this kind of synthesis cannot take place in a vacuum. A parent must be available to stimulate the child.

The following are examples of four ways of dealing with a likely interaction between parent and child concerning a situation the child finds interesting. The examples progress from least desirable to most desirable and illustrate how parental reaction can influence the inquisitiveness of the child.

> The child asks, "What's this furry stuff growing on the inside of my Jack-o-Lantern?"

> **Parent Answer #1:** "Yuck—don't touch that! It's rotten and smelly and we're going to have to throw it away. Run inside and get me a plastic garbage bag."

This is a lost opportunity to mention what a fungus is. What has been communicated is that the black furry stuff is yucky and smells bad. The child has also learned that her mother didn't want to deal with her powers of observation, and that she would be dispatched to get a garbage bag if she encountered more of the stuff.

Parent Answer #2: "It's mold."

In this case, the child has merely learned the name of the black furry stuff.

Parent Answer #3: "It's mold. Have you ever noticed it growing on old bread or on something old in the refrigerator?"

Now the child has learned the name, plus her parent has helped her recall seeing it before. She still doesn't know anything about it.

Parent Answer #4: "I see what you're looking at. It's called fungus, or mold. Get your magnifying glass, and you and I can get a better look at it."

This child is well on her way to learning something about fungus. She has, in the meantime, learned that her parent is interested in what she has discovered and sees enough value in it to stop what he's doing and come over to view it.

Below are a few tips that can help parents stimulate and maintain this natural curiosity about things natural and scientific.

- Remember that objects and events that are perceived to be important to adults kindle the most interest in children.
- Answer spontaneous questions and investigate that area as far as the child wants to go.
- Provide puzzling objects. This allows children to investigate the unfamiliar. Remember that unfamiliar objects tend to stimulate curiosity more than the familiar.
- Expose children to discrepant events and objects. This challenges children and generates curiosity, since children seem to view these items and events as puzzles to be solved. Children are most likely to be curious and question events that surprise them (this is

especially true of older children). An example of one such item would be a two-way mirror.
- Always investigate together in an atmosphere which conveys a feeling of "no-fail." Secure children are more curious than anxious ones.
- Provide children with pictures, television programs, and (when old enough) motion pictures with science content.
- Give them plenty of access to books, magazines, and other sources of information on topics in which they have demonstrated curiosity. Even such unexpected sources as seed catalogs, supply catalogs, and state park department magazines provide a wealth of interesting pictures, as well as information.
- Give children discovery boxes to keep treasures in. Help them identify the items and also investigate how they happened to be there.
- Let them watch the weather report at night and decide what to wear the next day.
- Give them magnifying glasses, and let them "teach" others how to use it.

Remember to make arithmetic and mathematics important in the home. Eventually, during the course of the child's formal science education, an understanding of mathematics will become imperative in the study of science. Not only will the formal operations of arithmetic be indispensable, the concepts of mathematics will need to be mastered in order to conquer such subjects as physics, chemistry, and statistics. These weighty disciplines may seem unimportant when dealing with the development of a preschooler, but it should be remembered that the ability to count develops around three-and-one-half to four years of age. At that time, children become fascinated with their new-found ability. This fascination can be intensified by parents who purposefully incorporate numbers into their children's home experience.

Parents need not do anything more specific than to allow children to practice number-related activities such as counting, measuring, and building.

Counting activities can include setting the table, allowing the child the opportunity to get the appropriate number of placemats, napkins—even pieces of flatware—from the drawer to the table. They can also include games, such as figuring out how many noses the family has altogether and how many more they have when Grandma and Grandpa come to visit. Early schoolers enjoy activities such as counting the number of raisins in a piece of raisin bread and keeping a record of the number in each piece of an entire loaf (this activity is especially fun when the children are allowed to *eat* the experiment!). There are many inexpensive toys that provide enjoyable opportunities for number theory practice. These include brightly colored cubes and pegs, games such as "Color-Shape Bingo," "Chutes and Ladders," and the traditional "Abacus."

Measuring can be practiced by allowing children to help cook the meals. Pouring liquids into different sized containers aids children in the development of the ability to understand volume. Comparing lengths, weights, and sizes of different objects promotes learning in the areas of geometry and spacial relationships. But very young children cannot be expected to understand too much in this area; it is tied up with intellectual development, as discussed in the chapter dealing with child development. Despite the lack of understanding, physical manipulation of physical properties, such as size and volume, still remains the only way these young children can gain any information about them.

Constructing things is a good method for children to apply both of the abilities discussed above, counting and measuring. Besides these, it allows them to practice the fine motor skills involved with the manipulation of tools.

Remember that if a child cannot seem to learn a particular concept, then it may simply be beyond her level of understanding. Sometimes no amount of description, hinting, or explanation will help her understand—she just can't because she hasn't developed that far. That doesn't mean that any experience she gains isn't knowledge; it's just not accompanied by the ability to fully understand.

Parents can greatly contribute to the science education of their children by interacting with them in science-related activities and experiences. These experiences will give the children a library of mental images, as well as an opportunity to flex their curiosity muscles. Questioning and teaching how to question helps children understand the principles and formulate a "big picture" of how all the puzzle pieces fit together. They gain perspective, and in the meantime, parent and child share a mutually enjoyable experience.

II

The Developing Mind

Infants and toddlers: learning through action

The development of mental processes in children is gradual, the passage from stage-to-stage being determined by a combination of forces, some more potent than others. By one process, children mature biologically and, as they get older, their intellectual ability progresses. In another maturation process, they interact with the physical environment and, as they explore into and interact with their environment, they gain information which promotes the ability to think. Finally, they gain considerable experience with the social environment, which exposes them to relationships and events associated with culture, cooperation, and competition. All of these processes combine to form the intellect.

The first two years of a child's life are a progression from the reflexes of a newborn to the beginning language of an eighteen-month-old. Development during this period is characterized by a concentration on learning through action. Objects are interesting because of what can be done with them. Events are interesting because of what an infant or toddler can see or hear during the event. Objects are not alive or dead, they just *are* what they *are*. As babies develop into toddlers, they begin to learn that they can have a direct effect on objects—that they can make things happen. During this developmental stage, babies and toddlers refine motor abilities, polish sensory perceptions and generally promote their own curiosity into the world of objects, actions, and events.

Newborn

The foundation of a child's personality is rooted in her early attachment to nurturing adults. Experts in early childhood development agree that the first requirement of good child-rearing is caring for the baby in a loving, tender, and attentive manner. At birth, all of the activities needed to sustain life are functioning; the baby can breathe, suck, swallow, and eliminate wastes. Newborns can also hear, see, taste, smell, feel, and signal for help which allows them to interact with their environment. However, in cases of infants deprived of that essential loving and attentive care, documented studies describe children who were unable to develop normally. It is loving care that assures a baby of progress and development.

The physical abilities of the newborn infant are reflex actions. Babies, only days old, turn toward startling sounds and stare at bright objects dangled eight to twelve inches from them. Parents naturally tend to promote these physical abilities by talking to their babies in the higher pitches known to be preferred by newborns, as well as coming very close to their babies' faces while talking to them.

After only four weeks of development, that same infant has become much more firm. Muscle tone is tighter, the twelve, tiny muscles that control the eyes are better organized and allow the infant to focus her eyes more often and with more control. At one month, faces and sounds elicit attempts to focus on the source, although the infants head is generally held to the right or left side due to the tonic neck reflex. Because of this orientation of the head to the side, interesting objects (such as mobiles) are better observed by the infant when they

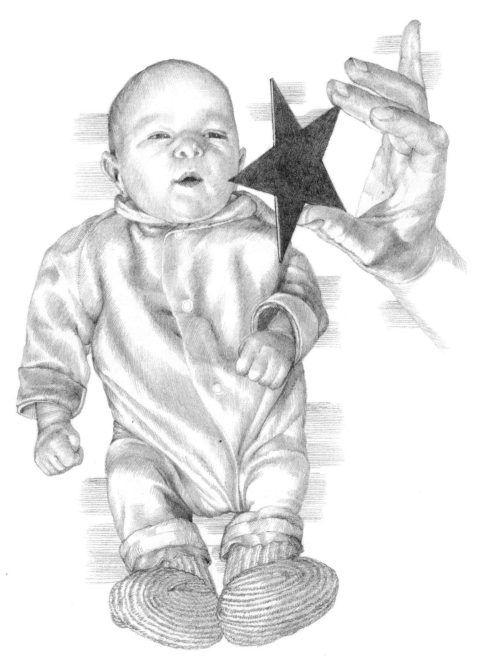

are situated on the side preferred by the infant. This stage offers greater opportunities for parent-child relations; the baby is awake for longer periods of time and is more alert and aware than immediately following birth.

4 Months

At four months, babies are awake for much longer periods of time and appear to enjoy looking around. The ability to hold her head up gives the infant a much wider vista. Given the opportunity, babies also appear to enjoy the new physical orientation of the sitting position since it also increases their field of vision. At four months, a baby's vision is equal to adult vision. The baby can see color and is able to focus on objects at different distances due to the development of the lens in the eye. Refined vision and focus prompts infants of this age to begin to reach out, and even though they are not yet effective at touching the objects that they see, they are, nevertheless, learning about three-dimensional space.

Accompanying the ability to reach out are the reflexes that cause babies to grasp objects, push at them, and put these objects into the mouth. These reflexes, in combination with an improvement in coordination, allow children to learn the properties of objects. Toys that can be handled and moved around stimulate these babies while, at the same time, they promote the ability to observe accurately—one of several skills vitally important to success in scientific endeavors later in life.

Observation provides the information that is learned through the senses: sight, smell, taste, touch, and hearing. From sight, children learn about color, size, location, and shape. From smell they learn to compare and identify smells and measure the strength of an odor. From taste, they learn bitter, sweet, salty, and sour. From touch, they learn texture and temperature. And from hearing, children learn pitch and amplitude as well as the sounds associated with specific objects and events. The ability to observe is the basis for all the other processes in science. It is also the first process that children use.

Eventually, by the age of seven, these children will be able to mentally recall the properties of objects without having to test them. They will have formulated an idea based on physical manipulation of objects during early development.

Four-month-old infants are verbal. They cannot yet speak words, but gurgle, laugh, coo, and babble. They can imitate several different tones and seem to spend much time experimenting with sound. They have an uncanny ability to locate a sound. They also recognize their mothers' voices, even when heard among many others. Her voice usually elicits lots of smiling, although four-month-old infants are big smilers in general.

At this stage, infants begin to become aware of themselves as separate from the outside world. They may even begin to show a preference for a favorite toy or object, a characteristic which shows an awareness that something exists outside of themselves. It is a good time to start taking them out of the nursery setting, exposing them to new people, places, animals, and objects in order to expand their world.

7 Months

The main concerns of babies at this age are objects, events, tones of voices, and other sounds. They are now able to sit up as well as move about by creeping or, in some cases, crawling. These infants are much better able to explore their physical surroundings by looking and feeling. Reaching out and grasping are performed with ease and, once holding an object, the infant is able to transfer the object from one hand to the other.

All of these newly refined skills add up to a more subtle ability to manipulate objects. The better the child is at manipulation, the more she can learn about objects and their properties. Seven-month-old infants are aware of different sizes of objects, such as blocks, and can compare them. They can discriminate between far and near objects and are aware of the space separating them. Of course, this is the age during which small objects are of great interest and danger.

Since the child is now proficient at reaching for an object and grasping it, her interest begins to shift to investigating what happens to an object when she does something with it. All parents have been frustrated by their baby's repeated act of dropping a spoon from the highchair. The goal of the child is not to frustrate the parent—it is to see what happens to the spoon when she drops it. Other similar investigations involve banging objects and throwing them, all in the interest of discovering the effect caused by a motor activity. This is the initial attempt to understand cause and effect. At this point, it is merely a physical action; but for these children, thinking is action. Therefore, they are learning. It follows that they need to be provided with a number of objects which are appropriately sized for dropping, banging, and throwing. The objects must be small enough to hold, but not so small as to be dangerous. In order that the child will not get an object stuck in her throat, it must be greater than one and one-half inches in any of its dimensions. It is important to also be aware of small parts of large objects that could break off.

Seven months is the age at which infants begin to find small mechanisms interesting, the kinds that involve cause and effect. Infants at this age find light switches interesting. They are not yet able to climb up on a chair and operate one, but they show interest in watching what happens to the light when someone else operates it. Jack-in-the-boxes are also popular due to their obvious "cause and effect" mechanism.

In her book, *Teaching Science in the Elementary School*, Donna M. Wolfinger describes the ability to determine cause and effect as one of the two main skills that underlie the important science processes of inference, prediction, and conclusion; the other main skill is the ability to recognize a system and its interactions. Although the ability to actually *determine* cause and effect in the interaction of a system does not come until the child's elementary school years, the acquisition of these abilities is known to be affected by experience as well as biological maturation.

It is not known to what extent experience with physical objects can speed up the process of intellectual development. But most science educators believe that the advantages of promoting this interaction far outweigh any possible disadvantages. That is why all the newer science curricula developed for instructing young children are heavily weighted in favor of physical manipulation of objects as the method of instruction.

Since the interest is present, it is appropriate for parents to promote the activities consistently performed by infants at this age. These activities are likely to be necessary to the infant's investigation into the properties of objects and the events they cause. If parents discourage these activities by, for example, consistently reprimanding the child for dropping or banging objects, the child may give up the investigations. This would tend to discourage the development of the ability to discover cause and effect.

Games that set up simple problems begin to gain the interest of these babies. Although they have not progressed to the point of looking for an object that cannot be seen, they do enjoy peek-a-boo and hide-and-seek games. They are also beginning to master the art of moving one object to get to the one they want.

Verbal ability is progressing. The infant is now able to make many different vowel and consonant sounds which will eventually lead to the first words. She may try to imitate sounds or sequences of sounds and may begin to use words, such as "dada" or "mama." It is most essential at this age that the parent talk to the child. Research has shown that many children are mentally processing their first words before the age of eight months, which means babies are listening to the vocabularies of the people around them. Conversations are best limited to very concrete, here-and-now situations. Talking to the infant about objects that she can see, touch, or hear will benefit the baby in her language development. This is also an appropriate age for story-telling, songs, and reading aloud. It is difficult to keep the child interested in a book or story for an extended period of time, but if the book or story is short and uncomplicated, it may serve in language development as well as a calming influence before sleep.

Like the four-month-old, these seven-month-old infants are fascinated by sound. They entertain themselves by practicing their own ability to make sounds with the mouth. They also love noisemaking toys such as rattles, bells, and music boxes. They may even attempt to imitate sounds, such as animal noises, or sequences of sounds as in adult language.

10 Months

At ten months, babies are able to crawl as well as pull themselves to a standing position. They can explore more thoroughly which once again increases the field of investigation. Greater dexterity allows the baby to probe and poke objects. Parents must prepare the house to make it as safe as possible for the baby without curbing curiosity and limiting exploration (see end of section).

At this stage of development, babies love to explore by "poking." The development of the refined ability to probe objects is due to the infant's ability to use an extended index finger. This allows babies much greater ability to investigate depth of space as well as texture. Fingers are becoming more important than the mouth in the exploration of object properties.

Babies at this age begin to show an awareness of top and bottom, container and contained, in addition to cause and effect. They relish emptying containers full of objects and then filling them again. They seem to be especially attracted to wastebaskets, gleefully crawling from bathroom to bedroom, emptying every basket.

These activities demonstrate that the child is beginning to develop one of the skills that will later be crucial to scientific investigation. The skill is classification, the ability to group objects according to their similar or dissimilar characteristics. The ability to classify begins with the ability to separate objects into two groups based on one obvious characteristic. The characteristic used to group objects during this stage is likely to be whether those objects are in or out of a container. This activity is only one of the first of many classification activities that the child will perform during her daily life experience. Nevertheless, it is appropriate to encourage it and name it for the child.

Stacking objects is now a favorite activity. Also, these children enjoy studying the movement of objects and may repeatedly roll a ball to the wall and back, just for the sake of watching it move.

A baby at ten months of age is able to reach behind herself to retrieve an object without looking for it. This action is the beginning of an understanding that objects are not necessarily gone when out of sight, referred to as the principle of object permanence. Understanding object permanence is an important step in the development of memory. Since acquisition of information through memory is the definition of knowledge, this development in intellectual ability is of obvious importance.

Infants may begin to build a one- or two-word vocabulary as early as ten months, although this spoken vocabulary is not a reliable measure of the number of words children understand. That number is much higher. It is, therefore, important to pay special attention to vocabulary. By this age, babies are investigating their own body parts; they demonstrate the acquisition of vocabulary by pointing to them when named. Likewise, they are becoming aware that objects around them also have names. Parents can help a child increase her vocabulary by simply supplying her with the correct word for an object or event at the time she finds it interesting.

Ten-month-old babies are exceedingly social. They enjoy having others around during playtime. Mutual games such as "pat-a-cake" and "peek-a-boo" are especially entertaining. These infants are developing their ability to learn by imitation; they may even modify their own manipulation of a toy after watching someone else's demonstration.

It is highly appropriate, at this time, for parents to begin to assess their own reactions to objects and events, realizing that the ability to learn by imitation will influence their child's interaction with those same objects and events. Showing, for instance, fear of storms or dislike of harmless insects, in the presence of the young child, will color her understanding of the situation. From now on, the child will watch her parents' reactions and learn from them. She will be developing lifelong internal reactions to objects and events, reactions that will be difficult to overcome later in her life.

This stage of development also initiates imitative behavior such as attempting to plug an electrical plug into the wall just as the parent was observed doing. It is difficult to react to these situations without eliciting some fear in the child's mind. Fear of an object is an effective deterrent to further investigation; but that deterrence is both good and bad. The child is safer if she refrains from investigating the plug again; she is also under the impression that she should be afraid of electricity. This impression may carry into her later years and prove to be detrimental to her tendency to want to investigate through experimentation. It is better to reprimand the child calmly and firmly. It is *best* to avoid the situation altogether. Hence, the following section on "accident proofing."

Accident Proofing

It is appropriate that a baby this age explore her environment.

However, many of the objects which children find interesting can be extremely dangerous. Since these children cannot yet comprehend an explanation of the dangers inherent in the exploration, the child needs to be carefully monitored when undertaking her "scientific research." She is modeling her parents' activities but needs supervision. So for at least a couple of years, the following suggestions are in order:

1. Move potentially dangerous items out of sight. Store them in an area that is out of the way of the young child. Secure those cabinets with a safety latch. This includes cleaning supplies, medicines, and personal hygiene products such as shampoo, hair spray, aftershave, first-aid kits, and talcum powder.
2. Never store a poisonous substance without a label. If the child should swallow it, you will know what it was. In most cases, these labels provide essential information about immediate handling of accidental ingestion. Post the Poison Control Number near the telephone in plain view.
3. Place houseplants out of reach whenever possible. Although some plants are notoriously dangerous when ingested, it is possible for any plant to be dangerous, depending on the plant part eaten. Pieces of leaves and small berries can cause choking, sharp pieces and thorns can cause cuts and scratches, and a number of physiologic reactions can be caused by eating a plant part according to that plant's level of toxicity. Reactions range from diarrhea to coma, or even death. Most books on treating childhood emergencies contain a section describing dangerous plants, but when in doubt, contact the local Poison Control Center or Botanical Garden for information.
4. Fill empty electrical sockets with plastic caps that are designed to prevent small children from removing them. Place furniture in front of electrical outlets whenever possible.
5. Keep all electrical appliances up and away, especially those generally used in the bathroom such as hair dryers and heated curlers. If a plugged-in appliance should fall or be pulled into water, electrocution could result. Electrocution is also a danger where an electrical device is used on a wet floor. (It is best to avoid using an electrical appliance in the bathroom when the baby is present.)

6. Replace frayed electrical cords immediately.
7. Fill bulb sockets with light bulbs.
8. Mount microwave ovens where it is impossible for the child to reach the controls.
9. Never leave cords dangling from countertops.
10. Keep trash baskets locked in child-secured cabinets or closets. Tie thin, plastic trash bags into knots before depositing them in the trash in order to prevent a child from putting them over her head and suffocating. Wrap any sharp or otherwise dangerous item before throwing it away.

12 Months

The one-year-old is on the brink of a dizzying number of developmental steps, such as walking unaided, building a tower, using a spoon, and speaking recognizable words. But these toddlers have made enormous strides since their tenth month in terms of their ability to explore and to understand.

These children are beginning to appreciate form as well as number. When given a choice between a round and a square hole, a one-year-old tends to thrust the extended index finger into the round hole. Despite the fact that they cannot quite yet build a tower, they do begin to orient one object above another object without actually placing the object on top. Given a group of cubes, these babies will place the cubes, one by one, onto the tabletop, which signifies the beginning of counting. The ability to count will later develop into the ability to use numbers to describe an activity or event—one of the important skills in scientific investigation.

At one year, children's interests generally fall into the following three categories: interest in their parents, refinement of motor skills, and exploration into the world around them. They are more likely to imitate than before, learning to master a skill such as scribbling after being given a demonstration. Games of give-and-take, such as rolling a ball back and forth from parent to child, improve their abilities to manipulate objects. Performances which draw laughter are much more likely to be repeated. Some commands, such as "give it to me" can be expected to be followed. Children at this age respond to music. They begin to appreciate rhythm and enjoy hearing children's songs. Also, most children are quick to "dance" for an audience of family members.

Perhaps the most important development at one year of age is the realization that parents are a source of aid and information. It is at this age that children establish a "social contract" with their parents. They learn how to gain the assistance of a parent during times of distress. They also learn specific answers to questions that develop during a normal day's exploration. By thirteen or fourteen months, a child has learned a great deal about her parent's style—whether her parent is the kind who gives undivided or divided attention, whether the parent will carry through on threats or merely "talk a big stick," and whether the parent is willing to allow an easy exchange with the child or not. She will also have learned whether her parent promotes or discourages investigation of objects.

Two of the most important social abilities that develop during the second twelve months of life are the ability to get and hold the attention of adults and the ability to use adults as resources when it is determined that the job cannot be done alone. These abilities have obvious implications for the parent who wishes to promote a child's tendency to investigate and question. First, it is important for the child to predict that the adult will give assistance when it is requested. Second, the child must learn how to gain that assistance. (Burton White, in *The First Three Years of Life*, warns that it is a disservice to the child to hover over her or pay too much attention to her, since she will not learn as much about the methods of getting and holding adult's attention.)

18 Months

Eighteen-month-old children have refined many skills. They are able to walk unaided, seat themselves in a small chair (with good aim) and even walk upstairs with a little help from a hand or a banister. They can build a tower with the newly developed ability to place one object on top of another. They can throw a ball and deftly turn the pages of a book.

Their intellectual development has expanded in many ways, having made great strides since the end of the first year. These children point to pictures of objects and know the names of them—some children may even say the names. They respond to simple commands such as "put the toy on the table." Instead of taking objects one by one, eighteen-month-old children enjoy holding a handful of objects, often attempting to use those objects to build a tower.

Due to their increased physical and mental abilities, these children are more thorough in their exploration into the properties of physical objects. Besides testing the action of the object, they begin to practice their own skills on the object. Spinning wheels are of major interest due to the action of the wheel. Water play is an especially favorite activity; small containers in the tub or sink will keep them busy, emptying and filling them.

Outdoor exploratory play becomes extremely interesting to this group. Leaves, dirt, sand, and insects are fascinating, and since the mouth has become less important than the hands and eyes in exploration, toddlers are fairly safe in their activities. Daily walks become slow-moving strolls as children spend more and more time stopping, looking, listening, and feeling. They appear to be thinking about their observations. The toddler often pauses for a minute during which a parent may even be able to predict his next acts based on facial expression or by the circumstances surrounding the child.

Questions are posed with inquisitive gestures or with puzzled frustration. Parents can answer those questions by demonstration since conversation is still not effective. An 18-month-old child may cry miserably while attempting to build a block tower on the plush carpet. The parent will alleviate frustration by building a tower of his own and starting it on a large book. In this way, the child begins to see that every tower needs a sturdy base in order to remain stable on uneven ground. The child's learning is strictly rooted in the actions she performs; she may only understand that her tower will stand up if she always begins it on that particular book. In her memory, though, is a picture of how a wide, sturdy base helped her stabilize a tower. It is important to remember to keep conversations about objects and events in the here and now. Children under two years of age are not yet able to think or talk about objects that are not present. All that exists for them are objects that can be seen, touched—experienced.

Preschoolers: asking questions

During the years from two to five, young children practice and perfect language skills. Although objects are still of primary importance, words are beginning to stand for those objects. The ability to talk propels preschoolers into a much wider interaction with people and surroundings, allowing them to gain the stimulation which is necessary for development of the intellect.

During these three years, children develop the abilities to describe and to explain, due to their increasing vocabulary and developing cognitive ability. These two abilities are directly related to the science processes of observation, classification, and communication. The level of understanding of science can increase if children are encouraged to practice these skills.

However, it is important to remember that these children think *much differently* than do adults in a number of specific ways. First of all, they see the world as revolving around themselves. It is impossible for them to take the point of view of another person, or to value ideas of another, which differ from their own.

They believe that no event happens by chance, and that each event that they perceive has something to do with them. There is no such thing as a random event. This mental immaturity is evident when children lash out at physical objects they encounter, such as blaming the "bad table" for stubbing a toe. As they begin to dispel this egocentric view of the world, they begin to ask "why?" questions. The reasons for events are no longer clear to them, although they still believe that there is a purpose to all events.

They begin, during this stage, to develop an intense curiosity about animals, due in part to the fact that they believe the animals share their feelings of sadness, happiness, fear, pain, love, and loyalty. Children of this age often project these same feelings onto inanimate objects such as cars, bicycles—even rocks and sticks. Preschoolers are so wrapped up in themselves that they believe all objects think and feel as they do.

In an attempt to understand their environment, these young children look for links between events even though they are hampered by their inability to separate real from imaginary, living from inanimate, possible from impossible. Events are often perceived as related simply because they follow each other in time, such as relating going to the hospital with having a baby (even when the person leaving for the hospital is Grandpa). Still, they are developing the ability to sequence events in time, learning that one event follows another event. This is an important process in scientific experimentation. Sequencing, along with cause and effect, helps to define a system and its variables.

Preschoolers are easily fooled by appearances. They lack the ability to understand that an object can remain the same even when its appearance changes. If the water from a tall, thin glass is poured into a short, fat glass, these children will preceive there to be less water based on the height of the water in the glass.

These children can classify objects according to one obvious characteristic. They focus on one property of an object at the exclusion of all others and cannot think of the parts and the whole at the same time. It is difficult for them to see oranges as being both round and orange; they may define fruits as all being round since the orange is round. These children base most of their understanding of things on visual cues, and the most outstanding impression of an object will determine their thinking about it. This aspect of thinking is related to the one described in the water glass situation. The higher water line proves the existence of more water.

They cannot link together a series of facts to form a general conclusion, nor can they travel through questions from point A to point B and then refer to point A. For example, when asked, "Do you have a sister?" a three-and-one-half year old answers "Yes." In response to "What is her name?"; "Her name is Betsy." But, in response to "Does Betsy have a sister?", she answers "No." Her understanding reached a roadblock when attempting to reverse her thinking to her own relationship to Betsy.

Finally, and most importantly in terms of understanding how children learn about the world around them, they must learn new concepts by using real objects. In order to learn about electric circuits, adults can read a book on the subject. Preschoolers must connect battery, wire, and bulb to learn anything about the same concept.

Preschoolers believe that there is a purpose for the existence of each object or event that they observe. This stems from their belief that all things are alive. By three years of age, children have begun to form their own notions about why events occur; these notions fall into predictable categories and any of them may be employed by the child to explain a curious situation. Individual children do tend to employ one category more often than others, although they do not use it exclusively. These categories of explanations are:

1. God causes things to happen (motivation).
2. Things happen just because they do (finalism).
3. Anything can cause anything to happen; water goes downhill because there are fish in it (phenomenism).
4. There is some sort of invisible force between two objects over a distance which allows one to act on another; the moon moves because the child is riding in the car (participation).
5. Things happen because there is a moral necessity; boats have to float because if they didn't, people would drown (moral).
6. People made everything and they can make anything happen (artificiatism).
7. Objects move and events happen because everything is alive (animism).
8. Life and forces are confused; mountains push the water and make it run downhill (dynamism).

These explanations are given up as the child grows biologically and develops intellectually. Research shows that it is difficult to actually teach children to replace these ideas with the true explanation for the causes of events; difficult, but not impossible. Children of four and five years of age tended to replace theories of animism, the belief that everything is alive, after learning the difference between living and nonliving through direct experience. In other words, by providing children with experiences that present both living and non-living objects, parents can help their children to understand that only living things eat, reproduce, and grow. When children understand this difference, they take a giant step toward a more sophisticated understanding of their environment. On the other hand, dynamism, or the mixing of life with force, is not given up after simply engaging in science experiences. Dynamism is a belief that children naturally take up in the absence of animism and it is one that they do not readily discard. It is the belief that, for instance, objects float because they are "light," because they have air in them, or because the water pushes them up. Children cannot understand the true reason why objects float until they develop a higher method of thinking, a development that occurs around the age of seven years which allows the child to consider more than one property of an object at the same time. Research has found that no amount of experience, direct or indirect, will cause the child to abandon dynamistic thinking until that child has progressed intellectually. Therefore, there will be, during these early years, questions to which the parent will not be able to offer an understandable explanation. In these cases, the best answer is generally an acknowledgement of the value of the child's curiosity along with a challenge to try to discover for herself. For example, in response to "Why does this float but not this?" the parent might say, "That's really interesting that you discovered the difference! Let's see if we can figure it out by trying some other objects."

Two Years

The potential for aiding a child in understanding the world of science and nature grows quickly once the child reaches two years of age. Many of the developmental traits described below lend themselves to the growth of a curious mind. Even some small traits, such as the child's refined ability to turn the pages of a book, to remain seated in a chair for a longer period of time and to remember events that took place yesterday, lend themselves to the child's interest in the exploration of the world and to the parents' ability to facilitate an understanding.

The two-year-old is becoming more aware of the world in general. Concentration is still focused on exploring the properties of objects and mastering motor skills on them, but these activities are lessening, along with an accompanying rise in the time spent watching people talk and listening to their language. The language can come from people, television, record players, or tapes.

Speech has grown tremendously during the passage of the second year. By the age of two years, a child's vocabulary may range from a few words to as many as 1,000, with an average of 300. Most of the words in this vocabulary are names of people, things, actions, and situations. Language is best taught at this stage by naming the object or event which, at the moment, is interesting to the child. Children of this age begin to ask the question "What's that?" and the potential for vocabulary building is unmistakable.

In asking "What's that?" the child has demonstrated that she has made an observation based on one or more of the senses. Observation is the basis for all the other processes in science, and once the child has observed something, the parent can reinforce the observation with vocabulary. Then the child has not only acquired knowledge through observation, she has gained a new word which will now allow her to communicate, or pass information, with another person. The two-year-old, after learning that the interesting object is a rock, will likely practice the acquired vocabulary by repeatedly asking "That a rock?" She is developing the skills of observation and communication.

Parents sense that during the third year, their children become thinkers. Their children begin to demonstrate reasoning powers; they will deliberately move a stool and climb up on it with the expressed purpose of obtaining an object that is out of their reach. Before this age, moving a stool and climbing it was simply performed for the sake of the action. As children progress from two to three years of age, they learn a great deal about properties of objects, cause and effect, and chains of events. They begin to be able to anticipate the consequences of some of these events. Since these children are becoming thinkers, these situations will gradually change from problems that are worked out with actions to situations that the children can think through before acting.

At this age, children begin to sense that anything that moves is alive. Therefore, an observed event may not mean the same thing to a child as it does to an adult. A tumbleweed blowing across a vacant lot has a different meaning for a two-year-old than for an adult, since the child's level of intellectual development leads her to the conclusion that the tumbleweed is alive because it is moving. Two-year-olds employ this animism to understand objects. Time and experience will cause them to give up this view of objects; until that development has taken place, no amount of explanation to the contrary will change their minds.

Two-year-olds are definitely beginning to show unique personality. They are beginning to use the pronouns mine, me, you, and I, although they are more apt to call themselves by their first names than by "me" or "I." Commands are more likely to be understood when phrased in this manner, such as "Mary jump down," rather than "You jump down." They show spontaneous affection for family members, laugh at very simple humor and even express some hints of pity, sympathy, and pride of accomplishment. They are able to distinguish the difference between bowel and bladder functions and may even verbalize the difference.

Along with the development of these positive traits comes an increase in negative behavior. This is an attempt to separate themselves from others, especially from their parents. Use of the word "mine" is on the increase, and its use is often accompanied by definite postures demonstrating the child's sense of ownership.

All of these traits demonstrate an increasing awareness of the child's own world. The two-year-old is physically as well as intellectually better able to explore and understand. Besides the development of language, reasoning, and personality, the child has become more flexible and has gained better balance. She can run, although she cannot yet stop quickly or make sudden sharp turns. She enjoys dancing, jumping, and clapping and accompanies much of it with laughter. Action and thinking remain closely linked. The child talks a great deal, even when there is no one to talk to. She can build a tower of six blocks as well as make a train, which requires the ability to line blocks up on a horizontal plane. She can use number words while she points and understands the concept of one versus many or more. During the course of a day, she may even be observed classifying objects in the following categories: hot-cold, big-little, soft-hard, and over-under. She may not always be correct in her classification, but she is beginning to enjoy the activity.

Three Years

Three years of age is a transitional stage, resembling in many ways the adolescent stage of the pre-teen. The child must move from the physical actions of infancy to the use of words as tools for understanding.

Three-year-olds begin to ask questions in their attempts to clarify their surroundings. In fact, at this age, most adult-child conversations come from questions and answers. These children are constantly attempting to identify and name objects, as well as classify and compare them, and they solicit help from their parents in their attempts. Therefore, questions of "What" and "Where" arise. The ability to learn by questioning is an important skill in science learning, as well as other areas of learning. Although these children are likely to ask *many* questions, they are not yet able to ask truly productive questions. It is important to remember, though, that these questions are asked in order to identify or clarify; it is extremely easy to overanswer children's questions, leaving them bewildered. Answers should be short and correct.

The vocabulary gained from receiving these answers is invaluable to the development of the child's scientific understanding. Words have taken on new meaning as the child begins to rely on them for describing ideas and perceptions. Along with the ability to express ideas with words comes an increased ability to form questions and carry on conversation. Both of these improve the child's communication skills and enhance the passage of information.

Three-year-old children still attempt to understand the physical and natural world by believing that everything can be alive based on the ability to move. As the child gets older, this definition of life will narrow, but at this stage it is very difficult to convince a child that something is not alive when the child believes it to be. These small children have significant fears. They often fear masks, "bogey men," old wrinkled people, the dark, and many animals. But attempting to convince these children to abandon their fears based on common sense arguments is frustrating. Objects that adults perceive as inanimate and, therefore harmless, may be perceived by the child as being alive and, therefore, threatening. Most of these fears will dissolve with experience, not discussion.

Three-year-olds are developing and refining a number of science-oriented processes which bear attention. They continue to be effective observers, watching events that occur during their daily experiences and storing information gathered through sight, as well as the other senses. By physically interacting with concrete objects, they add to the mental framework needed later on in order to be able to infer the properties of an object without actually coming in contact with it. They are better communicators, asking more questions and questioning more aspects of an object than just the name. This skill allows them to gain more information and also to describe the events that they find interesting. They are beginning to compare, and the ability to compare is essential to classification. In the course of a normal day, three-year-olds can be observed separating short and long, as well as more and less. They are beginning to recognize and name basic shapes and also primary and secondary colors. They may even begin to show an understanding of numerical concepts at the most elementary level, describing sets of one, two, and three.

In short, three-year-olds are learning to observe, describe, and compare. They will master these skills before they begin to explain.

Four Years

Four-year-olds have command of their language. They have many ideas and also have the ability to talk about them. They have the ability to ask many questions, and, to the chagrin of many parents, they exercise the ability with regularity. Unlike the three-years-old who asked the name of everything, the four-year-old wishes to gain information which will help to generalize the events happening around her.

These are the children who begin to want an explanation for the world surrounding them. Their questions are "Why?" and "How?" and are asked in an attempt to give order to their world. They have busy minds that bounce from idea to idea, often with very little elucidation. They may not even be interested in understanding the answer to a question; they may be waiting to see how the answer fits into their already established scheme. Often their questions are unproductive, such as "Why is the moon round?" These questions are merely a sign that they are attempting to assign a purpose to everything they observe. Their thinking is very literal, and for that reason analogies are difficult for them to understand. Although they are capable of carrying on a long conversation, they may not wish to repeat an idea in a discussion, saying "We already did that."

The most important thing for four-year-old children is that they be made to feel safe to pursue the answer to a question, safe meaning free from fear of failure or evaluation. It is, therefore, important to remember that the four-year-old views the world much differently than an adult, and much of this understanding cannot be altered by a mere explanation. These children have a method of understanding objects in which the ability of an object to move is the clue to whether or not it is alive. [See "Preschoolers: asking questions" for description of reasons children use for explaining events.] This perception colors all of their understandings of the world of nature and physics, and very little can be done at this age to alter that perception. Therefore, in order to maintain a "safe" questioning environment, children at this age should be allowed to view the world in this manner. Explanations that are contrary to this belief can be and should be offered, but the parent shouldn't expect the child to understand.

They are now practiced observers, and the further development of their fine motor skills lends itself to the use of a magnifying glass. With a lens, a whole new vista can be offered for observation. Vocabulary has grown allowing better communication, although it remains essential to remember that these children still learn best by coming into contact with an object or event. They do not learn what they cannot observe with the senses.

They are better able to compare and are now sophisticated enough to compare, for example, pictures of animals with their babies and match the correct mother and offspring. They can classify objects based on more than one property, such as separating all objects that are round as well as bumpy. Gradients such as small-smaller-smallest may be employed in their descriptions of sets of objects, rather than merely big vs. small. Four-year-old children show more interest and understanding in number relations, counting groups as large as ten. By encouraging their children to count objects, then pictures, and then drawings, parents can even help to introduce the number concept.

In summary, four-year-olds are verbal, social, inquisitive, and creative. Their explanation of events often mixes fact and fiction; in fact, they are not yet able to separate truth and fable. They enjoy the company of others outside their home, as well as that of their immediate family. They may have some unreasonable fears of such things as the dark, old men, and feathers, but their fears are fewer in number. They are bursting with activity, enjoying jumping, skipping, hopping, and climbing. They are striving to identify themselves with their surroundings while they peek into the unknown parts of their environment.

Early schoolers: forming ideas

During the first years of elementary school children enter the stage during which the egocentricity and concreteness of the preschool years gives way to the more mature thinking of an adult. Children begin to be able to consider perspectives other than their own. They value input from adults as they begin to doubt some of their previously held notions. Logic is developing and, although it is still sometime in coming, it affects the child's understanding of the world. Ideas which before were crystal clear (albeit, incorrect), such as the likelihood that all things are alive and have the ability to know and feel, begin to crumble in the face of this early logic, leaving children with a very unclear perception of how things work. They continually encounter situations that cannot be easily explained and, in order to replace old theories, ask adults for help in understanding. Thus, there is an onslaught of questions which are asked for the sole purpose of gaining information and understanding.

After the age of seven, children develop true logical thinking and the ability to reason. At that point they are no longer totally limited to the concrete world of objects and are on their way to true adult thinking. They are also becoming able to answer many of their own questions through reason, as well as through written reference. Until that time, they need a lot of help from adults, especially parents.

Five Years

Five-year-olds have developed a sense of self. This maturity makes them more comfortable with their surroundings and gives them more composure. They now appear to think before they speak, a dramatic change from the tendency of four-year-olds to blurt out questions and comments. Five-year-olds are more likely to seek the aid of an adult, both in questioning and in action. They show pride in accomplishment and enjoy being told they are doing a good job, making it even more essential, at this age, to bolster inquisitiveness and curiosity.

Questions are now asked for the sole purpose of gaining a better understanding. Five-year-olds ask questions because they want to know the answer; younger children often question for conversation purposes or for practicing speech. "Who made this?", "What makes this go?", and "How does this work?" are more frequent because children of this age are interested in practicality. Most of their questions deal with familiar objects and events, although an interesting television program or movie may spark questions about far-away places or unusual events.

Vocabulary is increasing at a dramatic rate. An average of two thousand new words have been added since the fourth birthday. Infant-type vocabulary is gone, and most five-year-olds are able to articulate all proper scientific vocabulary supplied to them. These children display a tremendous ability to remember past events. This ability, coupled with the tendency to ask many questions, allows five-year-olds to greatly increase their understanding of science and nature, as well as other aspects of the world surrounding them.

Unfortunately, there is a tendency to overestimate the ability of these children to truly understand the things adults have told them, since they can repeat back much of what they are told. Despite the fact that their vocabulary and syntax have developed, some intellectual skills have not. Many children at this age still embrace the concept of animism, whereby many objects are alive and have a purpose, even such inanimate objects as clouds and the wind. Concepts that they are able to talk about may still be beyond their grasp due to their intellectual immaturity. Parents need only remember to allow children to have these beliefs, since trying to rationalize a change would not be effective. Time, biological maturation, and experience will change these theories. In the meantime, in answering their children's questions, parents will have given the children something tangible even though their understanding may be superficial.

Some of the fears of four-year-olds have begun to disappear by the fifth birthday. Fears of bogey men, older people with weathered faces, and animals are being replaced by fear of the elements, including thunder, lightning, and heavy rains. These fears are more substantial when they occur at night, since they are joined by a fear of the dark. The questions generated by these fears are asked in hopes of gaining some comfort through understanding. Things that are understood are less likely to cause fear. In fact, a parent's calm, soft-spoken explanation can have a calming effect on a child. Even though she may understand little of the basis for thunder and lightning, she may find it less fearful if she senses that it does not cause fear in her parents. Often adults react to bulletins describing severe storm warnings or tornado watches with very obvious nervous or fearful mannerisms, such as rushing around for candles or ordering the family to the basement without explanation. Certainly there are times when precautions must be taken, but it is encouraging to the child to hear a calm, matter-of-fact tone from her parents. Young children learn a great deal from the cues their parents give.

Ideas are beginning to be accepted through rationalization; for example, a five-year-old may be convinced to try a new vegetable because of its vitamin content. She may hate it, but she will initially be more conducive to trying it.

At this age, children are beginning to grasp the finality of death. They describe it as a state without the abilities possessed by living things; when things are dead, they cannot move, talk, see, or eat. They understand that elderly people are the most likely to die, and they often mention that when they themselves get old, many others who are now alive will be dead. This demonstrates that these children understand something about the eventuality of death, although they have not yet grasped the idea that they will one day die. Still, they love to play dead.

Five-year-olds are aware of sex differences between males and females, and they know that the differences are indicated in sex organs, although they may not ask many questions about sexuality at this age. Boys and girls are still differentiated by clothing and haircuts. Babies are extremely interesting, but a four-year-old is more likely to question their growth and delivery than a five-year-old. They are more concerned with modeling the styles and activities of the men and women around them and may even begin to reject certain activities as "girl stuff" or "boy's stuff."

It is important to note here that by the age of five, girls and boys may already prefer to play with different types of toys. There remains disagreement as to whether this difference is inherent or whether parents contribute to it by choosing different toys for their boys and girls and by playing with their children differently. But despite this argument, it is a fact that in adolescence, girls tend to have a more difficult time with some math and science-related subjects, most notably those dealing with spacial arrangements such as geometry and trigonometry. In *How to Encourage Girls in Math and Science*, the authors suggest that little girls need to be encouraged to play with toys and games which involve spacing, visualizing, and grouping objects. These include blocks, balls, simple machines, and models; toys which historically, at least, were designated as "boy toys." In playing with these types of toys, all children develop special skills which will later be employed in many types of problem-solving. Parents need not encourage this type of play to the exclusion of other play which promotes verbal skills and interpersonal relationships. It is simply felt that girls may need a bit more encouragement in the areas of spacial orientation in order not to be at a disadvantage later on.

When five-year-olds do ask questions relating to sexuality and human development, parents must resist the temptation to color their answers with allegories of "seeds" or "eggs." To these children, seeds and eggs mean vegetables and breakfast. They need short, straightforward answers using scientifically correct vocabulary.

Science-oriented processes of observation, communication and classification continue to develop. Observation skills can now be sharpened through encouragement to use all the senses. Children can be asked "What is this?" "How does it look, feel, smell, taste, or sound?" And, since five-year-olds are developing the ability to analyze as well, they can be encouraged to broaden their investigation. "What is it doing?" "Why is it here?" "Have you ever seen it, or anything like it, before?" They are able to expand their numerical and spacial relations to include simple methods of measurement, such as stepping off the width of a room or measuring the width of a book by lining up and counting pennies. Even though formal measurement is not yet appropriate at age five, using these informal methods of measurement helps to build a basic understanding of the concept of measurement.

Six Years

Six-year-olds are much like five-year-olds in their tendencies to wander about and investigate the world of objects and events. However, children of this age are more likely to see an investigation to its conclusion, demonstrated in the fact that they not only enjoy taking things apart, but also putting them back together. They work well with simple tools, such as saws and hammers, and are actually able to build uncomplicated structures.

These children are very active physically and play best outside where they can create and invent using water, mud, and sand. They may even begin to show an interest in collecting objects from outside such as rocks, nuts, leaves, or even insects. They may show little fear of animals and are certainly old enough and discriminating enough to learn to predict which animals are safe for handling and which are not.

Discussions of past events and activities are becoming more meaningful as the strictly "here and now" orientation gives way to the less time-restricted orientation of the six-year-old. Although it has not yet been abandoned completely, the notion of animism is beginning to be replaced by a more correct view of the "aliveness" of objects in the universe. These children, based on their "hands-on" experience with the physical properties of objects, begin to accept more readily the real reasons why clouds move and wind blows. All the previously held notions explaining why events happen have not yet been given up, but by seven most will have been replaced. Children who have enjoyed much experience with the environment are, by this age, quite sophisticated in their understanding of the difference between random events and controlled events. Children raised on the farm understand the correct connection between the weather and the economy of the farm. They understand that rain encourages the growth of more grass that will feed the cattle meant to be used as another source of food. Experience is a good teacher and has a great effect on the child's tendency to give up theories such as animism. It also encourages the development of a true understanding of cause and effect.

Along with their developing ability to reason, these children gain the ability to ask and to answer the "What if . . . ?" question. For many

children, this ability to conceptualize will not develop until they are seven or eight, but it is observed in some six-year-olds. This ability greatly promotes the acquisition of scientific knowledge. It also allows the parent to suggest new avenues of investigation through the use of a thought-generating question. Once children become used to answering these kinds of "What if . . . ?" questions, they ask them of themselves; they may even ask their parents. A useful example of this type of question would be "What if it were raining outside?" "What would the sky look like then?" The parent might ask this in addition to answering a question about why the sky is blue. It nudges the child into a comparison of the appearance of the sky under two different sets of conditions. Of course, if a child has not yet gained the ability to conceptualize to this degree, it will not help her to understand better by expecting an answer.

Six-year-olds are excellent at classification, and they find the activity enjoyable. They can be encouraged to classify such objects as money. It is appropriate, since it can be related to daily experiences. (It can be given as an allowance, using different coin combinations adding up to the same amount.) This encourages classification by size and color and also encourages number relations. Six-year-olds, depending on their level of experience, may even be sophisticated enough to classify animals into groups of mammals, birds, reptiles, amphibians, and fish.

Number relations can be encouraged in many ways for these children. Most important for later science skills is the ability to describe an event in a numerical fashion. Such seemingly unscientific activities as playing on the see-saw are excellent for practicing this skill. A parent can develop an opportunity by saying, for example, "If you three want to play on the see-saw, you'll have to have two on one end and two on the other. So, you ought to go find another child to play," or "How many of you have to get on the see-saw to lift me up?"

Seven Years

At seven years of age, children are very verbal, but are equally good listeners. They have a great need for adult approval and are more likely than ever to come to an adult with a question. They are more aware of the broader–scale aspects of the world, such as weather, the heavens, nations, and the earth itself. They enjoy books, songs, T.V., and movies. Stories that capture their interest are more factual in nature, including stories about animals, space, and machines. Seven-year-olds are collectors, evidence of their developing classification skills. They are also readers. This is, therefore, a prime time to begin looking up an answer when a question is asked.

These children are able to make a good guess about the outcome of an experiment. They have generally given up animism; they now understand the true meaning of living and non-living. To them, only animals "feel" and "know." Objects are referred to as either "animals" or "things." Plants may or may not be categorized as something other than a "thing" based on the children's experience with them. Inventing things is quite enjoyable, especially when children are allowed to tinker with simple machines and even harmless electrical wiring equipment. Investigations such as these help to overcome any apprehension that may still exist as a carryover from the safety precautions stressed by parents at an earlier age. The meaning of electricity, for example, becomes clearer, and precautions make more sense to the child. Honored flights of fancy such as fairies, magic, and even Santa Claus now are met with skepticism, and their existence may be questioned in an almost scientific manner.

In short, seven-year-olds are more logical thinkers. They are developing a more adult form of thinking and can now understand some abstract theories, although their powers of abstract reasoning are still limited.

III

Areas of Interest

Human Biology

Q: "Why does my nose have two holes in it?" asks the three-year-old.
A: Air goes into the inside of your body through those holes and gives your body something it needs to work correctly, called oxygen. Air that has given up all its oxygen can go back out through those holes when it is finished. You also smell and sneeze through those holes—we call them nostrils.

Children consider themselves to be the objects that everything else revolves around. It is easy to understand why they find their own bodies and body functions to be so interesting. Body, in this case, means anything having to do with the physical self, its features, its movements, its aches and pains, its hots and colds—everything. Many things that occur in the body of a child may be happening for the first time, or the child may be noticing it for the first time. Whatever they can learn about themselves can be interesting, fun, and most of all reassuring. Whenever their physical functions can be related to other living and non-living parts of their environment, children can further understand general principles that occur over and over in the realm of science and nature.

This is not a book of anatomy and physiology. Rather than present a dry textbook walk through the human body, this book will give a child's eye view of the body and child-oriented explanations of it using concrete examples that occur in their daily lives. It will highlight the things that children wonder about the most, and whenever possible will relate those things to other animals, plants, and machines. Parents can easily pass knowledge of the human body, its internal systems, its growth and development, and its capacity to do work onto their children when they are able to predict the children's level of understanding and develop answers to the questions that are appropriately child-oriented.

The most important thing to remember about this section is that it always takes the position that the human body is an incredible machine, and that anytime a child expresses an interest in it, she is showing a positive inclination to want to discover more about science. If she is made to feel that any aspect of her body is bad or that it is not grounds for discussion, she will close a little door onto the world of discovery and understanding.

This discussion of the human body will begin on the outside and move in. Outward body appearance may become important to a child when she begins to notice the things about her that are the same or different from her friends. Also, it's a rare parent that can avoid comparing his child to other children—does she weigh as much, is she taller, does she have as much hair, is she more agile. Everyone is exactly alike in some ways; everyone is also very different. Some people are tall, some are chubby, some can run fast, some can see better than others. Everything about a person can be the same and yet unique; children need to be helped to feel good about themselves.

Skin and Hair

Q: Why do people have fingerprints?
A: All of those little lines, or grooves, in the skin of fingertips help fingers to hold onto things. Since every single person is just a little bit different, the lines on everybody's fingers are a little bit different. We call them fingerprints when we stamp a picture of them.

Skin is the thing that holds everything in. It serves as a watertight container that holds moisture in and keeps dirt, bacteria, and sun rays out. This is important since people are more than 50% water. It also helps to regulate internal body temperature through such mechanisms as sweating, goose bumps, and heat loss. It provides a way to sense the environment, since it contains thousands of nerve endings; luckily, people can feel through it.

Skin affects the internal climate with a lot of different structures and activities. One of the ways it does this is by having hair on it. Children often wonder about the existence of hair and why it seems to congregate in certain areas. Since hair is right there for the child to see and touch, its existence generates a lot of questions and it is very appropriate to talk about the nature of hair. Hair insulates, although in comparison with other mammals, people do not have a great deal of hair. It still does an important job. There is a large amount of hair on the head because it is exposed to the elements, and because the head is home to the brain, which is very sensitive to changes in temperature. Hair also protects the body openings from debris; thus people have nose hair, ear hair, eyelashes, and eyebrows, and, in men, facial hair. The hair under the arms and around the genitals is there to collect odors which at one point in time were very important in the lives of human beings. The final thing hair does is provide a means of sensing changes, since every hair is connected to a nerve.

Humans are relatively hairless when compared to the rest of the mammals. Skin tans in order to protect it from harmful sun rays. A chemical coloring called "melanin" builds up in skin cells in response to exposure to these ultraviolet rays. The dark chemical protects the cells from death, but if those cells are overexposed to the sun before melanin has built up, they will burn and die.

There is a reason why skin comes in different colors. People who are native to countries that receive more constant sunlight have dark skin to protect them from the harmful effects of the sunlight. People with fair skins are native to countries where there is much less sunlight and where the light is weaker. There is not as much need for protection, and pale skin is better able to absorb Vitamin D, one of the benefits of sunlight. One of the ways that skin controls body temperature is through perspiration. When perspiration evaporates, the skin is cooled. An example is found in the chilling affect of stepping out of the shower in the cool bathroom. Along with perspiration usually comes a flushing of the skin. This redness is caused by an increase in blood circulation in the skin. The closer the blood comes to the surface, the better able it is to give up excess heat. When the body temperature gets uncomfortably high, such as during a soccer game in June, blood rushes to the skin which is simultaneously being cooled through the evaporation of perspiration.

Skin has many different surfaces, each with a specific function. The cornea on the eye protects the eye and allows images to pass through, since it is transparent. The funny-looking folded skin on the elbows, knuckles, and knees is there to provide extra room when the joint is bent. Some spots that receive a great deal of wear, like the soles of the feet, get very thick. The ridges of finger-and-toe-prints are there to provide extra gripping power. Nails are the human versions of claws and at one time were indispensable for picking up and cutting. The openings into the body are lined with smooth, moist skin which provides a gradual transition from the outside to inside.

Skin provides protection, but it is also thin enough to allow sensations to be felt through it. By touching, people are able to feel heat, cold, dry, rough, soft, fuzzy, slimy, sticky, and painful. As every child eventually comes to find out, skin does have blood in it. Although the very outermost layer is made up of dead skin cell, immediately below that layer is a layer of living tissue full of nerves and blood vessels.

What Parents Can Do . . .

Comparing skin on different areas of the body is fun and informative, especially when it is done with a magnifying glass. Children are usually quite surprised at the intricate striation of the skin as well as the number of hairs.

Bones

> **Q: What would I look like if I didn't have any bones?**
> A: You wouldn't have anything to hold you up. You would be like a blob instead of being big and tall.

If skin is a sack, then bones are the things that hold everything up. Many other animals do without bones by having their skeletons on the outside (insects) or by living in a shell (clams), or by simply letting the water determine their body shape (jellyfish). But people have skeletons; they're the things that bodies are put on. Children have interesting bones in that many of them aren't hard. In fact, when babies are born, all of their bones are soft like the end of the nose. This rubbery substance is cartilage, and it is the material that the body begins to coat with minerals as the child grows. This is one reason babies need to drink milk; calcium phosphate found in milk makes bones harden in a process called calcification.

Even though bones are covered with a lot of fat and muscle, only a few of them are obvious, and children notice them either in passing or when they bump one of them into the coffee table. The knee cap, shinbone, pelvis, elbow, collar bone, jaw bone, and knuckles are very obvious and are great places for inspection of what bones feel like and how they move.

What Parents Can Do . . .

Let children investigate dried chicken or beef bones. Some very large bones can usually be obtained from the butcher. These are better suited for children under five who might still be tempted to chew on a smaller bone. Ask the butcher to cut a bone vertically, allowing the child to see the blood-producing areas inside the bone.

Muscles

> **Q: Why do people lift weights?**
> A: They want to make their muscles stronger. Muscles are the

things inside you that move your whole body and all the parts of you.

Muscles move the bones. They also move a lot of other things like the tongue, the heart, the lungs, and the eyelids. Muscles do more than just move things, for while they are moving and working so hard, they are giving off heat which keeps the inside of the body warm.

Children are not especially interested in muscles, and may never ask anything about them. This is probably because young children take movement and activity for granted. It often seems as though children only slow down to sleep, and even then they toss and tumble their way to exhaustion. They don't notice muscles because they never think twice about their own ability to move.

Bones stick out and they hurt when you bump them, but most children don't imagine bumping a muscle or breaking one. There are a few things that kids do tend to notice, like feeling rubbery-legged after running around the block or having to put a heavy load down that they have been trying to carry. What they have done is worked a group of muscles to exhaustion, and the muscle just cannot continue to move without some rest. One way to prove to a child that she is made up of a lot of muscles is to stand with her in front of the mirror and make faces. Making faces is not only a lot of fun, it's good exercise for muscles. No other animals make as many faces as a human can make because people have many more small muscles in their faces than do other animals.

Sometimes children notice muscles in the meat they are eating. A friend once related a story of just such an incident. She had prepared fried chicken wings for her three children, put them on the table, and then ran to the other room to answer a phone call. When she returned, her children were all giggling hysterically at their ability to "make wings flap," accompanied by a great deal of chicken sounds. It had occurred to them, quite by accident, that those joints, muscles, and tendons were there for a reason, and they had set out to discover what sort of movement they would produce. Lo and behold, the movement was flapping and, at that moment, the children had learned an important piece of information about chicken wings and the system of joints, muscles, and tendons. This is a perfect example of how an otherwise innocent incident can lead to scientific insight.

Muscles move bones in the following manner. Muscles work in pairs. For every muscle that moves something one direction, there is an opposite muscle to move it back. The muscles are arranged so they can act as levers for moving the bones. As on a seesaw, levers are good ways to move a big weight by using a small force. That is how a very large book can be lifted by moving only the hand and forearm with the elbow acting as a fulcrum. Tendons are white, stringy cords that allow a thick muscle to be firmly attached to one specific spot on a bone. Imagine how thick and unwieldy the fingers would be if they had to have a lot of muscles in them. Luckily long tendons extend from the fingers back into the muscles of the palm and the wrist, allowing fingers to be long and thin and strong. Flexibility is a measure of the ability to flex the tendons. Some people are more flexible than others, and younger people are generally more flexible than older people. But even children are more flexible in some places, less in others, explaining why some of them can stretch a little farther in ballet or gymnastics than others.

What Parents Can Do . . .

Try the "chicken wing" experience. Often young children will prefer to use cooked chicken, since it isn't so slippery.

Blood

Q: Why is the blood from my cut so warm?
A: The inside of your body is very warm. The blood got warm inside of you.

Every child is fascinated by his own blood; not a lot of blood—that's scary. It's the little cut or scrape oozing blood that gets his full attention. It usually hurts, and he's usually afraid that it will hurt even more when it gets washed. But despite all the negative issues involved, this kind of incident is one of his only chances to see the infamous red fluid that flows through his body. Its appearance generates wonder as well as apprehension.

The most important thing about the vascular system is that it has a master pump controlling the movement—the heart. It's difficult for a young child to imagine that there is really something inside the chest that's about the size of his fist that squeezes and relaxes over and over again, day or night, awake or asleep. It can't be seen, but there are many outward signs that the heart is there and that it's working. Those signs can be used to help a child understand that he does have a heart.

Many times children misunderstand the origin of the blood observed on the surface of the skin. To them, the blood seems to grow on the skin; its unlikely that they even associate the fluid with something inside the skin. One way to explain bleeding to children is to compare piercing the skin with popping a hole in a water balloon—the liquid inside comes out when it finds an opening. Of course, explanations similar to this one often lead to incorrect assumptions, as the author has found. After receiving this explanation, a child once explained that the blood coming out of his cut was just like when ice cream drips out of the bottom of an ice cream cone, only blood tastes salty. This child must have had an image of blood sloshing around all over inside of her like red Kool-aid, just waiting to drop out when a hole appeared. Not a bad deduction for a four-year-old based on the information given to her. The child may also be confused by the fact that blood doesn't naturally seep from body openings like the mouth, ears, or nose. Still, a child under four deserves some sort of explanation to her questions, and this one will suffice, temporarily.

Actually, blood moves all over the body in neat and tidy blood vessels that are like straws, tubes, and hoses. Children can relate to this kind of analogy because of the evidence they are able to see. They can see some of the blood vessels on the insides of their arms, right at the wrist and inside the bend of the elbow, or even better on the underside of the tongue. When you show these to your child, she will say "I wondered what those blue lines were. They're sure not blood! They're blue and blood is red."

The blue lines are veins; the tubes which constitute the pathway back to the heart and lungs. They are the ones which can be seen through the skin. In the veins, blood moves more slowly, the walls of the veins are thinner and they lie closer to the surface of the body; all of those characteristics contribute to the fact that blood in the veins can be seen.

But blood is only red when there is oxygen in it. Oxygenated blood travels out to the tissues through arteries which lie deeper within the body and which have thick muscular walls. They do not show through the surface of the skin. Tiny blood vessels, called capillaries, are responsible for allowing the passage of blood from arteries to veins. It is in the capillaries that oxygen is given up to the body tissues—capillaries are also the vessels which are severed when the skin is cut. When blood is seen on the surface of the skin, it is always red, since the moment it comes into contact with the air outside of the body, blood picks up oxygen and exhibits the dark red color associated with oxygenated blood.

Blood in the arteries is full of energy being pushed by the heart with great force. Each time blood is pushed out of the heart, more is accepted back from the body for oxygenation. For that reason, the heart is a pump. The pulse is evidence of the strength of the heart's push. It can be felt in the arteries, which are mostly buried deep inside of the body for protection—when one is severed, a great deal of blood is lost because it is pushed out with such great strength by the heart muscle. Children can feel the pulse in a few places where arteries come close to the skin, called pulse points. Two pulse points are the inside of the wrist below the thumb and alongside the trachea in the throat. By looking under the tongue with a mirror, children can see an artery visible as a thick pink line. Capillaries can be seen by pulling out the fold under the eye; many red capillaries traverse this area.

Blood pressure is often regularly taken in routine pediatric exams. It is frightening to children when the cuff tightens on the arm, especially when they are given no explanation as to what is happening. Blood pressure is actually very easy for them to understand. It is not only a measure of the push of the heart, but also of the heart's ability to relax after each push. Children accept the explanation that the blood pressure cuff is testing to see how hard the blood inside the arm can push on it.

What Parents Can Do . . .

By the age of three, a child should be allowed to watch what the lab technician does with a blood sample taken during a routine exam, if the doctor will allow it. Also, most physicians, when asked, will gladly explain to a young child the workings of the vascular system. Most are quite good at explaining in "children's terms," and an explanation can go a long way toward dispelling fears when a child has received a cut or a broken bone.

Breathing

> **Q: Can Tommy hold his breath until he dies?**
> A: No one can hold his breath until he dies; he can hold his breath for a while, but pretty soon he will faint and then he'll automatically breathe again.

One of the definitions of an animal is that it requires oxygen in order to be able to convert food into useful energy. Mammals, the group of animals that people belong to, use an incredible amount of oxygen because they have to maintain such a high body temperature. Lungs are there to pump the air containing the oxygen and also to provide a way of getting oxygen into the blood. Lungs also serve as the collecting tanks for the exhaust people make, called carbon dioxide. The specialized method of keeping good, oxygenated air from mixing with the old, carbon dioxide exhaust is called "breathing."

Once again, children are skeptical of anything they can't see, and no matter how far they can see down someone's throat, they cannot see lungs. Even though they can see the chest rise and fall during breathing, it's the rib cage they feel, not the lungs. It would be a real leap of faith for a young child of three to believe that inside her chest were two things like balloons, each one the size of a Nerf football, blowing up when she breathes in and deflating when she blows out.

There are a few demonstrations that prove the existence of lungs. It may be that a child will notice these things on her own and ask about them. She may wonder why the milk bubbled out of the cup when she followed her big brother's suggestion that she blow through the straw instead of sucking on it. She may wonder where the air comes from that makes bubbles in her milk when she tries that trick slowly. It may suddenly occur to her that she blows on her food to cool it in the same way she blows out a candle on her birthday cake. Or it may become obvious that something her parent is doing with the mouth is making the balloon expand. It may be as simple as wondering why she can float in the swimming pool. All of these questions refer to the existence of containers inside the body that hold air. These internal containers not only hold air that allows a child to perform all of the above-mentioned tricks. They also hold the oxygen that carries through the vascular system, providing energy to all parts of the body.

At some point, the child who has been helped to understand the functions of her body will begin to relate bodily functions to her environment. One of the most important interrelationships between body and environment is the breathing of air.

For a long time it doesn't occur to children that they are breathing. They don't have to think about breathing. As it is a reflex, they just do it. Eventually though, as children come to realize more about themselves, they become more aware of their reflex actions. When they breathe in, air rushes into their lungs. When they breathe out, they feel air coming out of the mouth. What they breathe in is mostly oxygen; what they breathe out is mostly carbon dioxide. They need oxygen to live. Since people are land animals, they get oxygen from the air. A fish or other water-dwelling animal might get its oxygen from the water. But, regardless of the source, all animals have to be able to get oxygen or else they cannot stay alive. Oxygen allows them to burn fuel within the body and when fuel is burned, the energy needed to remain alive is produced. Another thing that is produced as a by-product is carbon dioxide.

When people talk about breathing they say that they breathe air. A child may wonder what happens to all of that carbon dioxide that animals constantly exhale into the atmosphere. It is never mentioned when talking about the composition of air. Oxygen and nitrogen are usually the only two gases that are mentioned. So where does the carbon dioxide go? Certainly discussions of breathing will include a question like this from a child who has seen an old movie in which someone is trapped in a cave and is running out of air.

Plants use carbon dioxide. Not only that, they discard oxygen as a waste product. It is a very good trade. Animals give plants carbon dioxide and plants give back oxygen in return. One of the reasons it feels so good to walk into a greenhouse or lush forest is that the air is full of oxygen. In fact, people started bringing plants to sick people soon after it was discovered that plants give off oxygen. That practice is now a custom, but it is rooted in scientific fact.

Plants do not use carbon dioxide in the same way that animals use oxygen. In fact, plants need oxygen just like animals do, but they need only a little bit of it—hardly enough to measure. They use much more carbon dioxide in a process called photosynthesis. The word actually means light (photo) production (synthesis), and it is used to describe the mechanism by which plants make their own food using sunlight, carbon dioxide, water, and a few minerals. Plants are the only living things that are able to turn light into food. Everything else that is alive depends, in some way, on plants to provide them with food.

What Parents Can Do . . .

Help the child observe the difference in breathing before and after exercise. Children over five enjoy counting the number of breaths along with the parent. Also, stress the danger inherent in playing with plastic bags or plastic wrap, or of entering unused refrigerators and other containers. All of these items give a great risk of smothering—the inability to breathe. Too often, parents assume that little children understand the meaning of "You won't be able to breathe." They may not understand at all if they don't understand what "breathe" means.

Noses

Q: Why do I have a nose?
A: It warms the air you breathe into it, and you also need it to smell and to sneeze.

For most children, noses are for smelling and sneezing. But the nose is the main passageway for breathing air into and out of the lungs. Children, when asked to take a deep breath, always open their mouths. Perhaps it's because they feel "mouth breath" more definitely. The air that is inhaled feels cool and dry in the throat, and a large amount can be taken in quickly. Kids also like the impressive breathing sounds they can make through the mouth; it's tough to make those sounds through the nose—they often come out sounding unpleasant.

Kids think noses are funny looking, they have to be wiped with Kleenex all the time, even when they're sore, and they hurt terribly when you bump them. Actually noses are there to warm and humidify the air before it reaches the delicate lungs. The warming is done by the increased length of passage and the humidifying is done by mucus. Mucus does an important job moistening the air, as well as filtering unwanteds from it, like smoke, dust, even small insects. People have very different noses, possibly based on their heritage. People from warm moist climates don't need a long nose to warm and moisten the air. People from very cold or hot and dry climates need long noses to either warm the air or humidify it.

Digestion

Q: Why do we chew food?
A: We chew it so that it will get soft and be mixed with saliva. Then it can slide down a tube to the stomach.

Living things must eat. They either must "eat" sunlight, carbon dioxide, and water, and then make their own food, as plants do, or they must eat the food produced by something else. The drive to get and eat food is strong enough to be considered one of the most important determinants of an organism's life.

Man eats food. He opens up his mouth, puts the food in, chews it up, swallows it, and sends it down a long food tube that eventually empties out through the anus. Children notice the food they eat until they swallow it; as far as they're concerned, it's gone after that point. Their interest picks up again when what's left of the food prepares to leave the body, but that section will be covered later. Fortunately, people no longer need to catch their food in order to eat it. Everything else is just about the same as it was fifty thousand years ago. Not only that, our systems are very similar to the digestive systems of other animals, so animals' eating and elimination habits are similar to man's.

It is likely that young children have a difficult time understanding what their parents mean when they talk about the stomach. It may seem as though food drops from the mouth to the stomach like a rock down a deep hole. And who knows what happens to the food then, since it can't be felt after it is swallowed. Parents connect the abdomen with the mouth only when their child complains of a stomachache on Halloween night. "Too much candy," they say. Children may view it as punishment for having had too good of a time.

Since young children do not easily understand things they cannot contact directly, they have difficulty grasping the idea that what goes into the mouth has any correlation with what they eliminate as waste products. It is possible, though, for a child of four or more to understand that food and drink are taken into the body and that they are swallowed for a reason. One of the easiest ways to explain this is through "hunger pains." Many children want to know why their stomach hurts or rumbles when they get hungry. Those feelings are very good signs that food that is chewed and swallowed goes to a place inside the body called the stomach, and when it is empty, it has to remind its person to eat something. Children love to learn about digestion; it involves one of their favorite pastimes—eating. They are already very good at eating, but they probably don't realize how much work the body must perform on food before any benefit can be gained.

The teeth have to chew it, and while the food is being chewed, juices called saliva are mixed with it. There's a special chemical in saliva, an enzyme, which helps break down all the starchy compounds in food. When the food is chewed up sufficiently, it is swallowed, which means it is pushed down the food tube, called the esophagus. The esophagus is located behind the windpipe in the throat, and with its muscular activity it can move food from the mouth to the stomach in seven seconds.

It is delightful to watch a six-year-old suddenly discover the true meaning of swallowing. She may immediately challenge her body to prove how strong its swallowing power is—can she swallow upside down or laying on her stomach? She will discover that she can indeed swallow food, regardless of her position in space because of the muscular action of the esophagus.

Once food gets to the stomach, which is above the waist and between the two sides of the rib cage, it is turned into a white, milky fluid through a combination of acids and agitation. It's rather like the agitation cycle of a washing machine, and it is accomplished by the strong stomach muscles. The acids used are very strong, strong enough to eat away the wall of the stomach if it weren't for the protective mucous lining coating the inside of the stomach.

The food mixture is passed little by little from the stomach into the small intestine through an opening at the bottom of the stomach. It is in the intestine that the valuable fuel chemicals in the food are absorbed and passed into the bloodstream. The chemicals seep through the lining of the small intestine and pass into the capillaries that flow there. Whatever is left after all the useful materials have been absorbed is moved on to the large intestine where it is progressively robbed of more and more water and is passed out as a fairly thick mass. All along the small and large intestine, rhythmic muscle contractions force the food down and around through the food tube. When rumbles are heard from the abdomen, it simply means the intestines are working hard to do their job. When a stomach growls, it does so because it hasn't had any food in it for about eight hours. Without food, there is only gas filling the stomach, and the regular muscle action of the stomach on the gas causes growls and hunger pains.

One of the scariest things that happens to children is vomiting and, unfortunately, it happens when they are already feeling bad. There are even instances when vomiting must be initiated with Syrup of Ipecac in order to bring up what shouldn't have been eaten. Talking about vomiting, what it is, and why it happens can make it a less unsettling experience, since children mostly fear what they don't understand.

The human body has many emergency systems that go into effect when something goes wrong. One such system provides a sudden surge of adrenalin that allows them to have extra speed and extra strength in the face of a crisis. Their eyes fill with tears in order to wash away an uninvited guest or bit of debris. And in the case of vomiting, the brain signals that something is wrong somewhere between the stomach and the end of the small intestine by giving one a feeling of nausea at the same time the stomach is being filled by backward contractions of the small intestine. Finally the windpipe is closed, the stomach relaxes, and strong abdominal contractions force the food up the esophagus and out of the mouth. The intestines have simply said

"No thanks. We don't want this stuff down here. We're sending it back!" Syrup of Ipecac causes this otherwise natural reaction and is useful in getting something dangerous back up out of the intestines.

Parents need very little convincing that waste elimination is an important function to young children. Five-year-olds find immeasurable humor in this function, but their interest needs to be understood as an awakening to the processes their bodies undertake. They want to know what that "stuff" really is, where it comes from, and why it smells funny.

Bowel movements are the removal of solid wastes through the anus by movement of the bowel. Urination is the elimination of urine from the body. Urine is a clean and healthy fluid that is 95% water, 5% urea (waste product from the breakdown of protein), and some otherwise useful body chemicals which were in overabundance. Urine is produced in the kidneys, two extremely important organs that are located on either side of the backbone, just above the waist. Kidneys have the life-saving ability to filter wastes from the blood. Blood passes into the kidneys by way of two large arteries, after which it passes into many other smaller vessels and is cleaned in very tiny tubes. The wastes, in the form of urine, collect in the center of the kidney and then leave the kidney and pass into the bladder. When the bladder begins to get uncomfortably full, like a water balloon whose surface begins to get tight, little nerves send a message to the brain that says it's about time to empty the bladder. This is accomplished by relaxing the muscle that closes the opening, allowing urine to pass out of the body. Although urine, when it is produced by the body, is a clean fluid, it becomes a haven for bacteria when it is released from the confines of the body. For that reason, it is important that children learn personal hygiene; when they are deeply into the "why" stage—around five— they have the ability to understand the origin and usefulness of urine, as well as its tendencies to promote chafing and infection when it is not properly cleaned from the body.

Teeth

Children can easily study one set of body parts—teeth. Teeth are in the mouth to tear off and grind up the food that children eat, but

unless teeth start to hurt, get loose, or fall out, children don't think much about them. Nor do they respond to warnings that they will not have all of their teeth when they get old unless they care for them now. Children do not believe they will ever get old. It wouldn't be appropriate to scare them into taking care of their teeth by threatening them with a trip to that mean dentist who gives shots in the mouth. The dentist would probably not appreciate being painted in such a gruesome pose. What can be done is to remind them that there are living things in the air that get in the mouth all the time. Usually, nothing comes of these invasive organisms because there is nothing to eat in the mouth. But when children don't brush their teeth, little bits of food stick to the teeth, and "mouth germs" are only too happy to pitch a tent where they can find a long-term source of food, like down in those valleys on the surface of a molar. Children are easily convinced that toothbrush and toothpaste are weapons in the war on "mouth germs," and this approach can shed a new light on brushing teeth.

What interests young children the most about their teeth is the fact that they will eventually fall out and be replaced by new ones. It's certainly a legitimate question to want to know why. Little children have small jaw bones that grow as the child grows. Adult-size teeth do not fit into child-size jaws; so, to allow little children to eat food other than milk before their jaws are big enough for adult teeth, they get a set of baby teeth. These teeth are called deciduous, which means to fall off and is the same word that people use to describe trees which lose their leaves in the fall. Between the ages of five and seven years, these "deciduous" teeth begin to fall out to make room for the permanent teeth, since the child's jaw bone has grown large enough to support adult teeth.

Nerves

Q: Do worms have brains?
A: They have a bundle of nerves in the end of the head that's like a brain. But they can't think or remember like you can.

Human nervous sytems are so highly developed that they are able to coordinate billions of nerve cells all over the body to carry out essential body functions, while at the same time they remain prepared to deal with major emergencies at the drop of a hat. All of this is done with very little specific conscious direction on the part of the person. Children are not likely to notice themselves thinking. They are more apt to notice nervous reactions like the doodlebug curling up into a ball when it is touched or the dog spontaneously scratching when the child scratches the dog's "tickle spot." All of these situations demonstrate an animal's ability to sense its surroundings. Higher animals like worms, fish, rabbits, and people have a "head" end that has a bundle of nerves generally referred to as a brain. The bigger the brain, the larger the animal's capacity to deal with a complex environment. Mammals are able to see, smell, hear, touch—they get a lot of clues that help in determining what is going on outside of their bodies. Mammals, the large group of animals to which man belongs, sense their internal and external environment through nerves. Nerves are slightly broken chains made up of tiny nerve cells that carry electrical messages from the brain to the rest of the body and back again. This is very convenient, since if the nerves in the lips send a quick message back to the brain that says the liquid in a cup was too hot, the brain can turn that message into useful action by directing the neck to pull the head back.

Children hear about brains, but have little understanding of them. The subject of brains is usually limited to Saturday morning cartoon shows, but occasionally a parent may advise his child to "Use your head" or say "Weren't you using your brain?" A four-year-old may simply answer "I don't know; what's a brain?" It's difficult to imagine what children understand about their brains. They know that they see with their eyes, hear with their ears, and speak with their mouth, but where do they think?

The brain is the part of a child that sees when her eyes are closed, hears when her ears are covered, and speaks without opening her mouth. It is the master control center located within the skull, or head bone. It controls crying, fear, love, anger, and disappointment. It controls feelings and the ability to physically feel. It is the control tower, and everyone's brain functions in a different manner. People can be smart in a lot of different ways, and that is one of the things that continues to make life on earth interesting and effective.

Reproduction

Q: Can Susie ever be a daddy?
A: No, she can be a mother someday, but girls can never be a father. Only boys can be fathers, and only girls can be mothers.

Living things have an overwhelmingly passionate drive to reproduce, a drive so important that many animal life cycles seem to be meaningless after the animal has ensured reproduction, and so it dies. In the world of nature, living things are born and develop so they can reproduce—it's as simple as that. A lot of things happen along the way that make organisms fill important niches or appear beautiful in the eyes of man, but the real drives, both mental and physical, are to stay alive and to reproduce. The fact that the reproductive drive is so strong merely ensures that living things will continue to be on earth.

Little children notice gender early in life. Even toddlers are able to distinguish boys and girls or mommies and daddies in their books when nothing more than hairstyles are different. As they begin to notice parts of the human anatomy that are different, either between girls and boys or, more importantly, between children and adults, they begin to wonder why those differences are there. Many parents are confronted with questions about the baby brother's penis and why their inquisitive daughter didn't get one of those when she was born. Children are just as likely to ask parents about parts of their adult anatomy that the children can't find on their own bodies. A little girl might want to know why she doesn't have breasts (or even why she can't wear makeup to preschool).

Children need to understand what it means to be a child, biologically speaking, and what changes take place when a child becomes an adult. At birth, people are either male or female, and remain that way for the rest of their lives. The body parts they will need to reproduce are there at birth, or at least the capacity for developing them is there. How disastrous would it be, though, if four-year-olds could make babies? That would not be an adaptation that would be successful. All of the females would die trying, and humans would find it difficult to continue to thrive. Therefore, people have inborn delay systems that keep them from reproducing until their bodies and minds are ready to handle the birth, growth, and development of a baby person.

Until the time of adolescence, children do not have the means of reproducing and feeding a newborn baby, nor do they have the physical desire. They pretend to, since they see adults dealing with babies or with sexuality. But, biologically speaking, until sometime after the age of ten, boys do not begin to produce sperm, girls do not release eggs, and neither has experienced the sex characteristics that make males and females biologically attractive to each other. When they become adolescents and plunge head-first into the adult world of male and female sexuality, they should already understand what is happening to them and why—that biologically speaking, their bodies and minds are being dramatically altered to prepare them to fulfill their drive to reproduce. Of course, people exist and think on a completely different plane from the rest of the animal kingdom. People's sexuality is fiercely tied to a number of strictly human feelings and emotions. Nevertheless, the human body has worked very hard to produce a reproductive system that will serve its designated function—to make more human beings.

What Parents Can Do . . .

There are many excellent books that deal with the subject of sexuality on levels appropriate for very young children. Parents are most likely to feel comfortable dealing with the subject if they have perused the available literature and picked a book that they like before the questions arise. Then there is a ready source of information that parent and child can share whenever necessary. Until that time, it is imperative that parents use appropriate vocabulary when naming the body parts, engendering a feeling in the child that those parts are just as normal as the nose. Words such as penis, testicles, vagina, and breast should be used from the very first with young children who are learning their own anatomy. Later on, they will have a vocabulary with which they can form questions.

Death

Q: Will my parakeet ever wake up?
A: No, he is dead. He can no longer move or eat or make any sound because he isn't alive anymore. We will miss him.

By the age of three, children wonder about life and death. As discussed earlier, children pass through recognized stages in their attempts to attribute life to objects. Inherent in their understanding of life, though, is an understanding of death. In their very early years, when they think everything is alive, they don't spend much time worrying about death. Even when children begin to narrow the category of "living" down to those things that move, they may believe dead things are asleep. Eventually, though, as children begin to ask "why" questions, rather than just "what" questions, they will ask questions about life and death.

Children of urban societies are at a real disadvantage in this case. Rural life offered children much more contact with real life and death situations. It was not uncommon for a child to come in contact with dead animals, especially since much of the meat eaten by the family was obtained by slaughtering the farm animals, an event young children were not barred from viewing. Birth was an event children were encouraged to view, since all hands were needed when the ewes began to "lamb" or the cows began to "calf." Urban children have very little direct contact with these kinds of events, and, therefore, when questions arise, they have little experience to fall back on.

The trouble with death is that dealing with it brings up a lot of fears for everyone, especially children. Parents are most likely to have to answer questions about death as it pertains to wild things. Children are very good at finding dead animals, like insects and baby birds, and the death of these objects isn't so likely to cause a lot of fear, since young children see them as easily replaceable in the world of nature. But what if a cherished family pet dies, or more importantly, if someone known to the child dies? Parents need to be prepared to deal with their child's thoughts about all of these events, even if the child is under the age of four.

Parents are usually afraid to talk to their young children about death. Many feel helpless when their children come to them with a question like "Why did the little bird leave the nest if he couldn't fly? Now he's dead," or "Will my caterpillar wake up and not be dead anymore?" Parents often avoid discussions about death because, as with many other subjects concerning science and nature, they don't want to appear confused or uncertain to their children. Regardless of how hard parents may try to "not tell" anything to their children about death and its inevitability and finality, though, children will continue to act out their own feelings about it. Many children's games involve dying and killing. There's a reason for that; by playing these games, children are able to make death come right out in the open, making it something other than the shadowy demon it often is in their fears.

No one likes to think about growing old and dying, but cells are programmed to make it happen. Biologically speaking, members of a species die and make room for the young ones. Life on earth is productive because the things that live on it are able to adapt to change. This happens in the production of young things, with their new gene combinations. The result of the sexual union of male and female organisms, whether plant or animal, is an offspring not exactly like anything that ever existed before. Some of these new organisms won't be very well suited to life in their environment, and, consequently, they won't be around for very long. But other new organisms will be very well suited and eventually they will, as Isaac Asimov says, "Crowd out the competition." The problem is, if all the older, unimproved organisms stayed around forever, there would be very little room for the new ones. For that reason, accidents and aging cause many organisms to die off, leaving room for the new improved models. The species whose cells are programmed to age and die are more successful at adapting to the changing environment. This is shown now by the fact that the long-lived sequoia trees and bristlecone pines are nearly extinct, while the short-lived rat is very successful and, undoubtedly, will survive all but the most devastating changes.

If that explanation seems terribly cold, it's because it took into consideration none of the emotions associated with death. Parents already know what their own feelings are. Children experience the same kinds of fears in the face of death. The discussion avoided the spirituality of death, for it is appropriate for the parents to understand their own feelings about that powerful element of death and then communicate it to their children. But biologically, it seems important for some things to die so the young ones might live. Children need to feel free to talk about death as well as life. Parents needn't attempt to give them a full understanding in one discussion; willingness and availability when a situation arises are enough. Understanding feelings and fears helps, as well as understanding the biological importance of these events.

Creation

Q: What color of hair did the first man have?
A: No one knows exactly what the first man looked like. He probably looked a lot like your father, except he didn't stand up quite as straight and had a little more hair. Many people believe he looked a little like the apes, except he was much smarter.

The answer to the question of how life began is tied up in the personal philosophy of the parent. It's impossible to say with certainty how life began, since no one was actually around to see it happen. It's also unlikely that young children will ask about creation point blank, although it may slip into a conversation when a parent least expects it, with questions such as "Was there ever just one bird?" or "What color of hair did the first man have?"

Most scientists, regardless of their religious beliefs, believe that life began when some of the complicated molecules existing in the prehistoric oceans became able to organize simpler molecules into a molecule that was just like themselves. No one knows for sure what that molecule looked like, but once it had the ability to make itself over again, "life" began and was able to continue. In this way, the gases and waters of the primitive earth were able to progress into the beginnings of life.

Obviously, children do not understand molecules and their synthesis. But they do understand what it means to take a box of blocks, spill them out on the floor, and then very carefully put them together to form a system—a house, a car, a barn, or a church. That's how the earth and life on it began. It's just that when the building took place, it took a very, very long time before the blocks could be put together to form an organism that we could call alive. Once it was formed, though, life went about getting better and better at surviving in its environment and ensuring that it would continue to survive by reproducing itself, often becoming something better than before.

Along the way, an animal developed that would soon become the most populous of all the large animals. This animal can now live on the ice caps, in the jungles, on the deserts—even can walk on the moon. He can communicate, using his voice, his gestures, and all of those tiny muscles in his face that no other animal has. He is very intelligent and his head proves it. He, or she, is man, and he is the culmination of all the billions of years that life has evolved on earth. Children enjoy comparing themselves to the other animals; children stand up tall with hands that are good for grasping and holding tight. They can talk; they know who they are. They can think about tomorrow and yesterday. They are the inheritors of man's intelligence. They are able to understand any number of different ideas about how man was created, and they are most comfortable with the one that their parents believe —at least when they are very small. Given the incredible ability of the human mind, in the future they will reflect on the origin of life and of man and form their own opinion based on the past and on the present.

Zoology

Q: Why does the rabbit have ears that stick way up?
A: That way he can hide way down in the grass and still keep his ears up to hear something coming.

Children want an understanding of the web of life and how it is possible that each and every animal has a respectable purpose for doing what it does. They find animals intrinsically interesting, whether those animals nuzzle or slither. Children have an uncommon kinship with animals. They tend to believe that animals have the same feelings that they have: happiness, sadness, fear, loyalty, love, enthusiasm, integrity. To them, animals share their feelings because animals understand them. Likewise, children like to believe that they understand animals. All of the basic drives that animals share in their struggle to survive are easy for children to understand because of their kinship with animals. Parents can use this understanding to answer their children's questions.

Q: What is an animal?
A: An animal is something that eats, grows, and has babies.

An animal is anything that is alive but isn't a plant. Of course, many children are confused about the meaning of alive, so it is best to describe the activities of animals. Animals seem to be more lively than plants; they eat, they move, they look out on the world much the same as people do (which is appropriate, since people are animals). One definition of an animal, according to a child, is one that Alan Devoe gives in his book *This Fascinating Animal World*. He quotes a little boy who explained the nature of animals in this way: "An animal is something you feel like talking to." This definition has been put to the test over and over again with children, and it has worked every time. It demonstrates that level at which children relate to animals. They want to know about the appearance of the animal, its behavior or its environment. They also want to know if it is safe or if it will hurt them. And often, they will want to keep an animal that they find.

As discussed earlier in the section on the development of the ability to understand, most children do not fully comprehend the true meaning of "animal" until the ages of four or five. But even before that time, their interest in animals is well developed. One of the most effective ways to answer their questions about the definition of animals, regardless of age, is to allow them to compare representatives from each of the following groups: animals, plants, and non-living objects. Suggested materials might be several live animals such as turtles, insects, fish, family pets, rocks, tools, and houseplants. After grouping the items (or classifying them), parent and child can together investigate the following questions:

- Which group is the group of animals?
- Which are plants?
- Which ones are not alive?
- How can you tell which ones are alive?
- If something dies, how is it different from when it was alive?

These questions will be answered to varying degrees according to the child's level of development. They are merely asked to generate thought and discussion and not for the purpose of obtaining a "correct" answer. Younger children (under five) may have very creative answers which, though "incorrect" will display the notions they have already developed.

The answers may be quite amusing. But if the child has developed to the point that she has rejected the idea that everything is alive and is ready to accept the real meaning of living vs. non-living, then she may be able to understand the following concepts:

- Animals are alive.
- Plants are alive (this will depend on her experience with plants).
- There are objects which are not alive and never have been alive.
- Animals move, grow, and reproduce (this is the appropriate place to use correct vocabulary).
- Plants grow, reproduce, and are green.

All of these concepts, which are integral parts of the answer to the question "What is an animal?" can be reinforced with simple activities such as taking a walk through the woods, building a terrarium, or visiting the zoo or botanical gardens.

Body Coverings

Q: Why are worms slippery?
A: They are slippery so they can slide around easily underground.

Around the age of three years, children begin to notice and comment on the physical appearance of animals. Even before that age, they seem to be acutely aware of differences in body covering, although their experiences are usually limited to animals they can touch—family pets, slow-moving insects, and garden worms.

Children notice things like fur, scales, and feathers, and can infer much information about an animal when they understand the importance of a body covering. Body coverings protect the insides of the animal, and need to be appropriate for all aspects of his environment. They also protect an animal by allowing him to sense pain, pressure, and temperature, and provide a covering that can selectively release moisture or retain it as needed by the animal. Body coverings sometimes provide protection through special adaptation such as camouflage and mimicry.

Most animals have a body covering that is an additional layer covering the skin. The skin is literally a bag that holds in the vulnerable body organs and protects them from loss of moisture. It keeps animal bodies at the appropriate temperature, and it provides the animal with a means of sensing pain and changes in temperature. About the only animal skins that are frequently observed by children are frog skin and lizard skin and, of course, the child's own skin.

Scales

Q: What are these little pieces stuck onto the fish's skin?
A: They are scales; they protect his skin and act like a racing suit, letting him swim quickly and smoothly.

Body coverings come in many different forms. Fish and snakes have scales, thin little plates that lay over and protect the skin. Scales are different from fur and feathers, but a child may wonder why fish have scales instead of fur, not just why fish have scales. Both scales and fur provide extra protection, but scales allow the fish to live successfully in his own environment; his scales are like a racing suit in that, besides protecting him, they also let him swim quickly and quietly which improves his food-getting and escape mechanisms. Fur would surely weigh him down and would certainly reduce his chances of survival in the water; therefore, he does not have fur. Snakes also have scales. They are, like lizards, reptiles. But lizards have no scales. Lizards need less protection for their skin since they walk on legs. Snakes move by pushing their entire body against the ground and scales not only provide extra protection, they give a snake the ability to be quick and quiet.

Feathers

Q: Why is there a feather poking out of your ski jacket?
A: The jacket is filled with down. That's the name for the tiny feathers underneath a bird's big feathers. They keep birds warm, so people put the feathers inside jackets to keep warm.

Birds have a covering of lightweight feathers. Feathers are structures that help birds fly and keep them warm. The large flat feathers covering a bird's wings, body, and tail are made up of interlocking parts that let scarcely any air pass through, allowing birds to fly. The fuzzy, down feathers on baby birds are also under the large feathers of adult birds, directly against the bird's body. They keep the bird's body temperature correct, even when the bird flies through cold air. Manufacturers use goose feathers and down to make pillows, quilts, and down jackets. Children often notice the small feathers that mysteriously appear on down jackets—they provide the same warmth for the owner of the jacket that they originally did for the bird that so graciously provided them.

Fur

Q: Why is so much of Rover's hair falling out?
A: The weather is getting warmer and he doesn't need quite as much hair to keep him warm. We say he is shedding.

Some animals have fur, a covering of soft, thick hair which insulates the animal against harsh weather conditions. Fur traps air against the body, and the layer of warm air is so effective at keeping the body a constant temperature that polar bears can dive into freezing artic water for their food and still survive. Children wonder why some dogs have long hair and some have short hair or why dogs shed their hair when the weather gets warm. Animals who survive are animals who adapt best to their surroundings. Animals such as dogs originally flourished in many different places in the world under just as many climactic conditions. Dogs that were bred to hunt water birds have oily undercoats, like polar bears, which protect their internal systems from prolonged stays in cold water. Dogs that originally were bred in cold climates have long, insulating hair. Dogs bred exclusively as indoor pets have hair that is attractive but does not shed. It's often possible to play a guessing game of "Where did this animal come from?" by examining its fur.

What Parents Can Do . . .

Diversity in body coverings is easily observed by looking at different groups of animals (such as animals in the zoo) or by collecting pictures of animals. Children understand best the differences among body coverings when they are allowed to feel the differences. When allowed to compare, for instance, a turtle's covering and a puppy's covering, it becomes obvious to children that there is a difference. They can then be asked questions such as:

- Does the puppy's fur protect it when the weather is cold?
- Do people wear coats that are like the puppy's fur? Why?
- What other kinds of coverings do animals have?
- What other animals have fur?

Camouflage

Q. Why does that bug look just like a leaf?
A. He looks so much like a leaf that it is hard to see him. That makes it harder for a lizard or bird to find him and eat him. He is safer. We can say that he is camouflaged.

Coloration is a specialized method of adapting to the surroundings. It's as if animals are wearing a disguise, and they use their disguises in many different ways.

One way that coloration protects is by allowing the animal's body covering to blend into the surroundings. This kind of protective coloration is called camouflage, or as many children like to think of it, the "G. I. Joe" method. Children have wonderful eyesight, but it is still unlikely that they will see a green lizard in the holly bushes until it moves. The eggs of the kildeer, a common shore bird, look so much like the color of a pebbly beach that they hardly need to be hidden; they simply cannot be discerned by human eyes, or by the eyes of any other threatening beast.

Q: Why does a fawn have spots?
A: Those spots make the fawn very hard to see when she lies still in the forest. They look like the light and dark spots you see when you look into the woods.

Many animals blend into their environment through optical illusion, although when they are viewed out of their normal habitat, such as in a zoo, this illusion is not effective.

Some snakes have many spots, stripes, and lines, and therefore they're easy to see when they're slithering across a dull piece of ground. It is nearly impossible to see them, though, when they lie still against a colored background. Fawns have many spots of white hair mixed into their brown coats to produce this same optical illusion since they are a source of food for many forest predators and need extra protection. The animal's markings produce an optical illusion that conceals his outline. Butterflies, although often brilliantly colored in flight, close up their wings and cling to tree limbs, exposing only a small area of wing, the area containing the least amount of color.

The ability to stay still is an effective method of protection when combined with camouflage. Animals that seem so full of energy can become completely stationary at a moment's notice and blend in. Baby birds are especially good at this; sometimes it's the only way for them to survive when they overestimate their ability to fly and prematurely leap from the nest landing in a backyard full of dogs and children!

Some animals combine camouflage and standing still with crouching down, thereby eliminating a shadow that might otherwise give them away. Children have always enjoyed games employing this hiding technique, and they enjoy knowing that animals hide that way, too. It's as if a moth were playing a hiding game when he presses his wings against the side of the tree trunk. He's actually trying to eliminate his shadow. Polar bears do the same thing, since their shadows would stick out like a sore thumb against the white snow —they crouch down against the ground. And little sand crabs run for the footprints when they feel threatened; despite their excellent protective coloration, they still can't eliminate the shadow without getting into one.

Q: Why are all of these fish white on their tummies?
A: That way, if a big fish looks up at them in the water, they will look light like the sunlight and be harder to see. The white tummy protects them. It is camouflage.

When the sun shines on an object, it brightens the surface it shines on and darkens the opposite side. Many animals have developed an adaptation that balances this shading effect by darkening their top sides and lightening their bottom sides. This adaptation is called countershading, and its effect is tremendously important to many small animals that are otherwise rather defenseless, like mice and birds. Fish show countershading, too. Their top sides are dark and their stomachs silvery. When viewed from above, they're difficult to see, because the water is dark. When viewed from underneath, their silvery undersides blend in with the light from the surface.

> **Q: Why does this bug look so much like a twig? I almost thought it was a twig until it moved!**
> A: Since it looks just like a twig, it is safer. Animals that want to eat it can't find it when it stands on a branch. When animals disguise themselves like that, it's kind of like wearing a costume.

There are animals that take on a resemblance to some other animal or object to hide themselves or to fool their prey. This adaptation can be described to young children as an animal wearing a costume. Although color and optical illusion remain greatly effective survival techniques, many animals simply take on a disguise. Moth and butterfly wings are very much like leaves, and when the animals are at rest with their wings folded up, they look like single leaves. One butterfly is actually named the dead-leaf butterfly because he so closely resembles his namesake. Many a child has marveled at the little green or brown leaves that hop on and off of his shirt out in the yard. These are appropriately named leaf hoppers, and even children easily understand how they protect themselves from being eaten.

Other animals resemble bark and twigs. Many moths look just like the bark upon which they sit. Some moth caterpillars, such as the measuring worm, are able to hold themselves in a precise position to allow them to look like a twig. Walking sticks and praying mantis look so much like twigs that they are often impossible to find. This resemblance to twigs isn't just for protection. Praying mantis are predators; they depend on their ability to catch insects for food, and the disguise makes them good hunters.

> **Q: Why does that bug have stripes like a wasp? Is it a wasp?**
> A: It isn't a wasp. It wants you to think it is, though, so you won't

touch it. The birds might be fooled, too, and not eat it. It is wearing a costume to protect itself.

Another form of protective resemblance is mimicry. Sometimes animals imitate the color, shape, and behavior of dangerous or offensive animals to fool their enemies or their prey. Many moths, beetles, and caterpillars take on wasp stripes without taking on any of the sting. It's a brave animal that will take a chance on eating these insects, knowing that they might be wasps. The robbery fly looks very much like the stinging bumblebee, and the king snake closely resembles the poisonous coral snake. The harmless animal mimics the dangerous animal and, therefore, enjoys the respect given to the dangerous cousin while retaining its own mild character. It's easy for children to understand why those flies that look so much like bees won't hurt them when they understand that the fly is wearing a bee costume so he'll look scary, just like a child wears a ghost costume and says "Boo!" on Halloween.

Q: Why did the chameleon turn brown? He was green when I caught him!

A: His color changed because he is in danger. You can tell that he is afraid and wants to go back to the bushes.

Some animals change color when they are angry or afraid. They give up their protective coloration when threatened and turn dark in an attempt to look fierce. Children are most likely to see this phenomenon when they try to catch an anole, or "American chameleon." His color changes when he is caught; the reptile is afraid and is attempting to scare his captor into putting him down. Easily observed phenomena, such as these, inevitably spark questions from a child. In the case of the anole, the child who understands why the animal changes color will respond to the animal's fear. He may identify with the reptile and let it go. He will invariably try to make the animal "feel better."

Q: Why does that wasp have black-and-yellow stripes?

A: He has those bright colors so that everything will be able to see him and remember to leave him alone because he stings.

A few animal species possess conspicuously bold and beautiful coloration. They seem to be saying "Here I am!" Their bold appearance isn't merely for beauty's sake. They are really saying "Here I am! I dare you to bother me!" The most beautiful animals are often the most dangerous, like the hornet and the coral snake. These animals are displaying their belief that they are free to be bold and conspicuous, since anything that tries to eat them will be making a fatal mistake. A good rule to give children is to avoid touching or bothering anything that is very conspicuous. That animal may indeed be saying to the child "I dare you!" Some animals, such as the monarch butterfly, don't harm children, but they are very distasteful to the birds that might feed on them. They fall under this category of warning coloration, and certainly they'll be pleased to have children refrain from picking them up. Of course, many animals are brilliantly colored in order that they might attract the opposite sex. Experience is the best teacher in this case; after years of observing animals and delving into their physical makeup and behavior, children will come to separate harmful from harmless. It remains a good rule of thumb, though, that if it is bold, do not touch it.

Body Parts

Q: Why does our cat have these soft pads on her feet?
A: They help her to creep up softly on something if she is trying to catch it.

Body parts are also adapted for successful survival in the environment. They provide the animal with help in searching for food by promoting quickness, agility, stealth, or strength. Cats have pads on the bottom of their paws that allow them to creep up on their prey. Eagles have large, broad wings that allow them to soar in the air, searching for their food.

Body parts may also be adapted for eating, such as the wide variety of teeth used for eating all sorts of different diets. Cows have large, flat teeth to grind their vegetarian diets, while wolves have long, sharp teeth better suited for ripping and tearing the flesh of a meat diet. People have both kinds of teeth since people eat plants and meat.

Finally, parts of the body may be adapted for greater protection.

Rabbits have extremely long ears that allow them to hear better; their ears stand up high enough to hear over the grass in which they live. Cats have retractable claws that provide them protection against their enemies while also enhancing their food-getting ability.

Animals adapt to ensure their survival in their surroundings. If some characteristic helps them either improve their food-getting or ensure their protection, that characteristic is going to be used by the successful members of the community and, therefore, will be passed on to the offspring.

Animal Behavior

Q: How do baby chicks know how to peck their way out of eggs?
A: When he is ready to hatch, the baby chick already has a picture in his head that tells him "This is how you get out." No one has to teach him; it's called an instinct.

There are two types of behavior, one that animals are born with, and one that they learn after having been alive for awhile. Inborn behaviors are: instinct, reflexes, and social behavior; they can't be easily changed. Learned behavior includes things such as trial and error, conditioning, and habits; these can be changed voluntarily.

Inborn behaviors, such as reflexes and instincts, are things that animals do simply because their brains tell them to do it. Baby ducks swim because it is an instinct. Chickens peck their way out of eggs by instinct. Migratory birds fly south for the winter because there is a message in their brain that tells them to. These kinds of behavior are often observed by children. But it is difficult for a child under five to understand that an animal does things for any other reason than because he wants to or because someone told him to. For this reason, it is difficult to try to explain instinct to young children. It is much better to tell them that baby chicks want to get out of their eggs to see their mothers or that baby ducks swim because their mothers taught them to, until they reach the age of four or five and are ready to accept a more difficult explanation. A child of five who has observed a great deal of animal behavior may begin to suspect that the behaviors she is seeing are not ordinary "learned" behavior. The fact that a chick can peck its way out of the egg or that a newborn foal can find its mother's milk will undoubtedly raise intriguing questions. When, at the age of four or five, a child begins to question these kinds of behavior, asking such insightful questions as "How did he know to do that?" or "When did his mother teach him that?", it is a signal that she is ready to talk about instinctive behavior. She may begin to understand that many animals are born with the directions on how to do a certain thing printed in their minds, as if they were born with a message in their brains that said "Now, do this."

Insects show more of these instincts than do birds, birds show more than dogs, dogs more than man, and man very few, if any. It is as if the lower animals are given an extra boost, since they are not as good at thinking as are the higher animals.

Animals living together in an organized manner, each member performing a task which is helpful to the group, are displaying social behavior. Bees and ants are the favorite examples, and the best way to explain their behavior is through a comparison with the society in which people live. Some members of an ant colony care for the eggs, some collect and store food, some maintain the tunnels and pathways, and some protect the other members from attack. That's not very much different from human society, although people don't perform their jobs by instinct. Nevertheless, ants are much the same as people to children, and as far as children are concerned, ants do things for the same reasons people do. Let explanations reflect this childlike view of nature.

Learned Behavior

Q: Will my puppy ever roll over like Tommy's dog?
A: We can teach her to sit up and roll over; she just needs lots of practice and special treats when she does it right.

There are, of course, many behaviors which are not inborn. Learned behaviors are those which come after practice and experience, like teaching a dog how to sit up or, better yet, toilet training a two-year-old. People are intelligent animals, meaning they have the greatest ability to learn new behavior. Animals which are the most like people, such as the apes, are more intelligent than the lower animals, such as the insects. That is why it is more difficult to teach an ant to do a trick than it is to teach a dog.

The most effective means of explaining the difference between learned and instinctive behaviors is in terms of teaching the animal. No one needs to teach a male dog to lift a leg in order to urinate on the fire hydrant, but it is necessary to teach him not to do it on the furniture.

What Parents Can Do . . .

The more animal behavior children observe, the better they will understand its purpose. For this reason, the best way to answer children's questions about animal behavior is to help them observe more of it. Habitat-oriented zoos are helpful, although special caution needs to be taken to separate "caged" behavior from "natural" behavior when observing animals in the zoo. Other helpful activities are building an ant farm, keeping an aquarium and hanging a bird feeder near a window low enough for a child to easily see out.

Ecology

Q: If all the animals outside are going to the bathroom all the time, why doesn't it smell bad?

A: They go in little bits all over the place. There are lots of bacteria, or germs, that clean it all up.

To a child, the world is like a big house. Ecology means, actually, the study of the house. It is the study of the relationships among animals and plants and their surroundings. Since every living thing, including people, must depend on other living things and the non-living things that surround them for survival, it's easy for children to relate to the predicaments of other animals. Children know what it means to be hot, tired, scared, sick, and lonesome—so do the animals. Children know how the elements can control these situations in their own lives. They want to know if the same situations affect their animal friends, the ones that live in their world— their "house."

Children between the ages of three and seven begin to notice that many of the things they do and many of the items they use during the course of a normal day are also done and used by animals. Animals drink water, eat food, live in particular places that they consider home, and have babies. Children question some of these shared behaviors and also wonder about the resources that they and their animal friends are using. Since these shared resources are an inextricable part of the "house" that all things live in, they are an appropriate starting point for a discussion of ecology.

Materials used by living things are returned to the system so they can be used again. Every time a material has been taken, used, and returned, it has gone through a cycle.

The food chain is a cycle of energy transformations, since whenever an organism uses food, the food is transformed into a form of energy that will fuel the animal's system. Some organisms make their own food from the nonliving products in their environment. These are the producers—the plants. They can use sunlight, carbon dioxide, and water to make their own food through the process called photosynthesis. The food they produce provides them with the energy they need to stay alive. It also provides energy for all the organisms that can't make their own food. Animals, and many single-cell organisms, can't make their own food from light, carbon dioxide, and water. They have to eat food to get the energy they need for life. That's why they're called consumers; they consume, or use, the food that the plants make. They do this either by eating plants or eating animals that eat plants. Animals that eat plants are called herbivores. They have large flat teeth in the back of their mouths for grinding up all those stems and kernels. Their front teeth are adapted for breaking off the plant tissue they eat. Grazers, the ones which eat grasses, have regular front teeth they use for biting off the grass leaves. Animals that eat bark and wood, such as beavers and rats, have long front teeth for gnawing. Other animals eat organisms that have eaten plants; they are carnivores, or meat-eaters. When they catch other consumers and eat them, they are called predators. The animals they eat are their prey.

Children are quick to notice this constant battle of "eat and be eaten." They enjoy watching herbivores chew on plants, such as the squirrel eating the sunflower seeds in the bird feeder. That is not a threatening situation. But predator-prey relationships are much more difficult for them to accept.

When a child's pet cat proudly presents him with a dead mouse, the animal is often viewed as being mean or cruel. The same is often thought of a snake that has captured a pack rat. The feelings of the child are brought on by feelings of empathy for the captured animal, as well as by many children's tales which tell of blood-thirsty wolves and terrible grizzly bears. Children feel much less threatened by these incidents when they understand that these animals must eat, and that in the mind of the animal, he is only doing what is needed to stay alive. People are no longer prey to wild animals, except in a few rare cases, so it is unfair to describe predators as mean or bad for capturing and killing animals. People are significant predators themselves.

The meat bought at the grocery store is muscle tissue taken from animals killed at packing houses. There is nothing wrong with that, just as there is nothing wrong with other animals killing for food. Meat is a significant source of protein, vitamins, fats, and energy for active animals. Besides, predators complete an important job in the community of animals. They keep the population down at a level where there is enough food to go around. Without predators, many animals would die of starvation. Cats and dogs have predatory ancestors similar to lions and wolves, and even though they are now domesticated and enjoy family life with their people, they retain a bit of the instinct of predators. That's good, actually. It might save them if they are ever lost and without food.

There is one more group of consumers that neither eats plants nor catches prey. They are the scavengers, and they get their food by eating animals that have already died. Few children have failed to notice buzzards crouching right in the middle of the highway feasting on last night's unlucky opossum or skunk.

What Parents Can Do . . .

One of the most effective ways to communicate with children about the food chain is to share with them the excellent nature-oriented programming on television, almost exclusively shown on public and cable stations. These programs, such as NOVA, NATURE, WILD AMERICA, and others offer humane approaches to what is often understood as an "eat or be eaten" world. Children generally find these programs fascinating. But a word of caution needs to be addressed. Often this type of programming includes graphic footage and audio of what are historically "scary" animals—bears, wolves, sharks, etc. Since these are often the animals of which nightmares are made (see the Child Development Section), parents should view these shows with their children, discussing the events and animals shown and addressing any fears the children express. Often, just watching and talking about these animals dispels many fears, especially when the parents seem comfortable.

Food Chains

Q: What do plants eat?
A: They make their own food inside their leaves with sunlight, air, and water. They produce all the food on earth.

Food chains are a means of transferring energy from organism to organism. A friend's child once asked "Why do plants eat at all? They don't need energy because they don't do anything." Plants do something; they grow.

Fortunately, for all the consumers and decomposers, plants produce a lot more food than they need to grow and then they store it. Consumers eat the stored food in plants to get energy to move, catch their food, and eat. Each time energy is passed from one organism to another, some is used in activity and some is lost as heat. Energy cannot be recycled, so it is imperative to all life on earth that we have the sun that continues to send limitless supplies of energy to earth every day. That's why the transfer of food energy is called a food chain, rather than a food cycle. That's also why all organisms on earth would perish without the sun.

Decay

Q: Why does that dead bird smell so bad?
A: There are bacteria on it that are turning the animal's dead body back into dirt and dust. When they work, they breathe out a kind of smelly air called gas. The job they are doing is called "decay."

If the producers make the food from sunlight energy and pass that food energy on to the consumers, how does the cycle become complete? How does the source of food return to the beginning, or recycle? If it is not recycled, will it be used up, ending the cycle of life? Luckily, there is another link in the food chain called the decomposers. It is their job to cause dead organisms to decay, and in their ranks are many fungi and bacteria. They have a very important position in the environment since they are the ones responsible for helping to return many of the materials needed to sustain life back to the environment. Without the decomposers, materials wouldn't be recycled and would eventually be used up, and the earth would be covered with a layer of dead organisms that had not decayed.

The most obvious thing about decay is the odor that often accompanies it. If a child brings in a dead frog from the sidewalk, it is apt to be recently deceased or long gone, beyond the smelly stage. She's not likely to pick it up when it's been there five or six days in the heat of the summer. When decomposers digest material that has protein in it, they produce bad-smelling gases. Children often understand this best when bacteria and fungi are described as having bad breath when they eat meat and animal products, such as meat, fish, eggs, and cheese. It certainly doesn't make the smell any more pleasant to know where it comes from, but it does make it more acceptable. It is also one more clue that there really are tiny little organisms on the decaying organism that do a specific job which directly benefits everyone. This knowledge can help your child understand what a refrigerator or freezer does—it retards or stops the growth of decomposers. A bowl of tuna salad has to be left in the refrigerator for a long time before it will smell even remotely similar to the same bowl left outside for two days in July.

Communities

Q: Do bugs live in bug cities?
A: Most of them don't, they live alone. But some do, like ants and bees.

Children find places like the lawn, the garden, and the shrub beds interesting because they support a lot of interesting organisms. A parent's help can go a long way toward promoting their understanding of relationships of living things. Parents can show their children that these places support little communities, just like their own community. Even scientists describe places where organisms live together as communities. A lawn is a community. It's a home for grass, dandelions, worms, insects, and even mushrooms. All of the organisms in a community depend on each other for sources of food, protection, and other things. A child can understand that many organisms call his home their home, too.

Even though organisms live together in communities, they don't all exist for their mutual benefit. Many organisms experience competition. Competition occurs when products needed for survival are in short supply. Children are apt to notice that the grass doesn't grow well under the magnolia tree in the summer. The tree and the grass are competing for sunlight, and the tree is winning. The concept of "survival of the fittest" comes from this competition factor. In the wild, the healthiest organisms survive, while the weak or unhealthy ones, or those unable to adapt to change, die.

What Parents Can Do . . .

Ant farms are tremendous teachers and are available through many hobby shops and some pet stores. They can be stored up and out of reach, and, therefore, investigation can be easily controlled by the parent, making them appropriate for even the youngest children.

Parasites

Q: Will the fleas make Rover sick?
A: They will make him itch when they bite, but since he is their home, they won't make him too sick.

There is another important relationship historically interesting to children because it affects their close companions—their pets. This is the parasite-host relationship in which one organism, the parasite, lives on another organism, the host. Although parasites stay alive by living on their host and feeding on him, the parasites never take very big bites. Understandably, this would not be in their best interest, since they depend on their host and prefer him to be alive. Children are most likely to come in contact with parasites when they notice ticks or fleas on their pets. Of course, it is quite possible for them to experience this relationship first-hand if they contract head lice or are bitten by fleas or ticks. All of these organisms are parasites, and they are simply living in a way that will allow them to stay alive.

When parasites are found on pets, most children want to know if the parasite will hurt the animal. Parasites in small numbers are a nuisance, and they have the ability to inadvertently pass on serious diseases. It is the large infestation, though, that threatens the well-being of the animal; for this reason, veterinarians advise pet owners to attempt to keep their pets free of parasites. The same situation is true for people.

Helpful Relationships

The natural world has many examples of helpful relationships. Birds build their nests in trees, and clownfish hide in the safety of the tenacles of the sea anemone. Ants keep tiny insects, called aphids, in their tunnels during the winter, providing the aphids shelter and food. The aphids, in turn, produce juices that feed the ants.

Dangerous Animals

Q: Is that snake poisonous?
A: Most snakes aren't poisonous, but before we try to pick it up, let's check in the snake book.

When parents answer their children's questions about the animal world, they are providing a sense of community with other organisms and their nonliving surroundings.

One other thing most parents would like to be able to tell their children is which animals are dangerous to touch and which animals are safe. There is no way to unquestionably classify an animal as safe or dangerous by the way it looks.

Much of the ability to know the difference comes from experience. As mentioned earlier, that's how animals learn what to avoid. But, there are a few rules that will help predict an animal's "danger factor" in an unknown situation. These rules can be built by adding

specific organisms, such as brown recluse and black widow spiders, rattlesnakes and cottonmouths, even poison ivy and nettles.

- Avoid animals that are brightly and beautifully colored; they may be daring you to touch them.
- Respect an animal's heritage. For example, if he's a wolf or looks like he's related to one, he may have a bit of the bite left in him (this goes for unintroduced dogs).
- Never try to separate a mother from her baby. She may have a method of protecting her offspring from whatever she perceives as dangerous predators (such as a parent or child) that is not pleasant.
- Don't bother an animal that is contained, but would rather not be. A dog on a chain or a bird in a cage may just be waiting for a hand or finger on which to take out all of its frustrations.

Temporary Guests

Q: Can I keep him?
A: Only if we can take care of him correctly, and we can only keep it for one day.

Once parents have shown an interest in their children's powers of observation of the natural world, they may be asked to play substitute mother or father to any number of different animals. As children get more comfortable with their environment and the organisms that share it, they begin to take a personal interest in some of the animals, personal to the point of bringing them inside. Parents may be asked to provide food and shelter to such varied local inhabitants as ants, caterpillars, lizards, crickets, worms, frogs, grasshoppers, toads, moths, snails, snakes, tadpoles, and turtles. If a parent is willing to accept a child's unique association with the other living things in his environment, he may also be willing to accept a temporary guest for the sake of closer observation. There are a few guidelines that can be followed when deciding whether or not an animal should be kept as a pet, even for a short time.

- If it is young, it should not willingly be taken from its mother.
- It shouldn't be kept if it cannot be made comfortable.

- If it's not used to captive life, it should only be kept one or two days.
- It should not be handled very much, and then only with great care.

These guidelines are ammunition to back up either a refusal to allow the animal to be caged or an insistence on setting one free.

When the decision is made to allow a child to keep the cricket he found under his inflatable swimming pool, he must be made aware of the things that are necessary in order to ensure the survival of a captured animal.

- It needs a cage with enough space to move around comfortably.
- Inside its cage, the habitat should look very much like the environment in which the animal lives in the wild.
- It must have a place in its cage to hide.
- It needs the right food, clean water, and plenty of ventilation.
- The cage must be kept clean and should not be allowed to get smelly.

The following is a list of foods that have been found (by Glenn Blough and Julius Schwartz, authors of *Elementary School Science and How to Teach It*) to be appropriate and palatable to a few popular wild animals. This will help the child provide good food for her guest.

- *Ants*   Dead insects, bread crumbs, ground nut meats, and water (place on top of soil).
- *Caterpillars*   Leaves that it was eating when found.
- *Chameleons*   Small, moving insects and green branches with their stems in water; sprinkle leaves with water.
- *Crayfish*   Chopped meat and water plants.
- *Crickets*   Fruit, lettuce, bread.
- *Earthworms*   Good, rich soil.
- *Frogs*   Earthworms, caterpillars, living insects.
- *Grasshoppers*   Leaves they were found eating, or celery and bananas.
- *Rabbits*   Various kinds of green vegetables; wild rabbits get water from the dew on grass.

- *Land Snails*   Lettuce, spinach, grapes, and apple.
- *Snakes*   Earthworms, insects, small pieces of meat wiggled in front of them. If they won't eat, let them go where you found them.
- *Toads*   Insects, earthworms.
- *Turtles*   Turtle food, insects, lettuce, earthworms; put food for a turtle on top of the water since many will only eat underwater.

What Parents Can Do . . .

Foster a feeling of camaraderie between your child and the animal that can include letting the animal go after a short time. Wild animals are seldom suited for living in a cage, and certainly are not happy there. There is justification for allowing your child to keep one for a day or so because he can observe it more closely. It also helps build a bond between your child and his animal neighbors when he realizes he can touch them and care for them, and then put them back where they really belong.

Animals and Seasons

Q: Do animals like snow?
A: Snow makes it harder to find food, so some animals leave during the winter. But some stay; they don't mind the snow—maybe they like it just like you do.

Young children sense the changes in the seasons. Plant life changes drastically in response to seasonal changes in temperature, humidity, and length of daylight. Animals, too, react to these changes, and their methods of dealing with seasonal changes differ greatly. Some of their adaptations are very obvious, and children wonder and ask about them. Some are not so drastic or are hidden, and these require a little extra help from an adult if a child is going to recognize them.

Young children have great difficulty understanding the actual reason for the changing of seasons. The system of planets and stars is far beyond their ability to understand; it has no physical object that the child can touch, few phenomena that she can actually observe. Therefore, it is safest to simply say that in the summer, there are more hours of sunlight, so the earth is warmer. There are fewer hours of sunlight in the colder seasons, so the earth doesn't have as much of a chance to get warm. Children can observe that, and it isn't incorrect. It is just an appropriately partial answer.

Spring is great; everything gets to work again. It's a time of renewal. Children begin to notice that the whole world is getting active again. Birds are showing up again that haven't been seen since last fall. Insects are beginning to be seen again, hopping around in the grass and flying in the air. Grownups are out cleaning the garage, something they seem to do every year about this time. Lawns that were dormant are beginning to look green again, maybe not with grass yet but certainly with weeds. Flowers bloom, and trees begin to bud. It's obvious to even the youngest naturalists that the warmth and the rain have sparked growth and activity. Seasonal changes have also turned everything's thoughts to "fancy"; the birds, bees, and just about everything else sets out to duplicate itself in the enjoyable job of reproduction.

Immediately following spring is summer, the season of plenty—plenty of food. The food supply is greater during the summer than any other time of the year. One of the things that determines animal and plant survival is the ability to make summer's food last all year. People are especially good at this. Fruits and vegetables are canned, frozen, or otherwise stored, grains are dried, and meat is smoked or salted. This is all done to make food produced in the summer last all year and to make it difficult for the competition (insects, rodents, worms, fungus, and bacteria) to eat it in storage. Plants also store summer's bounty. Structures such as potatoes are produced during the summer so that the plant can make it through the winter when the tender top has been lost.

Summer inevitably fades into fall, the season during which everything prepares to last through the winter. Some animals move to a more hospitable climate; birds migrate South, and children love to spot the ducks and geese as they pass overhead on their way to a warmer place. Their migration is rather like going to the Caribbean for Christmas vacation. By November, most of the migratory birds have gone; the birds that are still around will be staying for the winter. The birds that stay for the winter are seed-eaters, not insect-eaters. Most insects die of the cold and all that remains during the winter are their eggs. The animals that do not migrate are either equipped with survival skills that permit them to stay active during winter's cold, or they hibernate.

Most of the cold-blooded animals go into hibernation in the late fall when the temperature begins to get really cool. Cold-blooded animals are reptiles, amphibians, worms, insects, spiders, and fish. (Fish are exceptions to this rule; they have an amazing capacity for withstanding cold and swim around under the ice all winter.) These animals are called cold-blooded because the insides of their bodies are subject to the same temperature fluctuations as their environment. For this reason, when it gets very cold, many of these animals go into a state of suspended animation since they are able to carry on few body functions when their temperature is so low. Frogs, toads, and turtles bury themselves in the mud and stay completely still until the weather warms up. That's why they seem to disappear during the winter and mysteriously reappear in the spring. Snakes hibernate too, but occasionally during a warm spell, they may be seen soaking up the warmth on top of a rock. They go back into hibernation when the respite from winter's cold has ended, and they don't come out for good until winter is over.

Warm-blooded animals either migrate, hibernate, or continue to feed through the winter. Mammals and birds are called warm-blooded because their body temperatures remain constant regardless of the temperature of their surroundings. This is accomplished by a number of different attributes including high rates of metabolism, production and storage of fatty deposits, and insulating body coverings like fur and feathers. Warm-blooded animals that are not adapted to cold weather, despite their body warmth, migrate to warmer climates where food is more plentiful. Some, like the bear, remain in their old climate and sleep all winter. These animals eat a great amount of food in the summer and build up much fatty tissue that helps to keep them warm while they sleep. Still others stay and eat all winter. These animals are either successful foragers, like the deer or elk, who are able to dig around for food and find enough to stay alive, or they are predators whose prey stay all winter. Competition is fierce, but many manage to stay alive. When winter finally arrives, everything stays relatively quiet. It's a peaceful season during which most animals are attempting to keep warm.

Seasons cause animals to change their behavior. Children can relate to this. Their play changes, and the food they eat changes, just like the animals. Their clothing changes just as the animals change their coats. Observation of seasonal changes is an effective way to help children relate their way of life to that of the animals.

What Parents Can Do . . .

Bird feeders are excellent sources of information about animals and seasons. Different birds populate the feeder at different times of the year, and when bird seed mixes are used, children will notice that different seeds are preferred depending on the visitors.

Botany

Q: Where did all of these stickers come from that are stuck to my socks?

A: There was a plant back there in the field that wanted you to move its seeds to a new place. They stuck to you, just like they would stick to a rabbit or a dog—many seeds take a ride on animals by sticking to them.

Even the youngest children are fascinated with flowers. Flowers have attractive colors and shapes, pleasant fragrances, and seem to be a haven for wandering animals such as butterflies, bees, and lady bugs. It doesn't occur to most children under the age of seven that plants are alive just like animals are alive. To young children, plants are just there. Since their thinking is based on the ability to actually feel or see something or to perform a physical act upon the object, they have difficulty imagining that plants eat, grow, develop, reproduce, and die—just like all the other living things in the world.

Children between the ages of three and five are most likely to notice obvious plant structures and changes in those structures. They will not, as a rule, inquire into the life cycles of plants, their food-producing mechanisms, their reproduction, or their below-ground structures. These questions are more likely to come from a five to seven-year-old. The questions of three to five-year-olds are more likely to involve flowers, leaves, thorns, buds, and plant structures people eat. As they become more familiar with plants, their questions will become more intuitive, but during their early years, they will simply wonder about the parts of plants that they can see.

Children take an interest in two kinds of plants—the ones used as a source of food and the ones with noticeable structures that attract attention, like flowers or thorns. As children begin to get more involved in the process of meal preparation and widen their eating patterns to include many different kinds of fruits and vegetables, they begin to notice some of the fascinating secrets that plant structures hold. They also become better at observing the intricate structure of flowers. As they are included in the caretaking of the lawn, flower beds, shrubs, and house plants, children start to question the usefulness of structures they observe. They wonder why roses have thorns or why grass blades are so sharp. Up until now, they were not allowed to touch the houseplants and probably weren't invited to help pull the weeds. Along with all of this horizon-broadening comes a great deal more thirst for information and explanation.

Despite the fact that children of this age group do not tend to ask about the physiology of plant growth and reproduction, many of their questions will inadvertently touch on these subjects. A question about the reason leaves are green must, in its answer, include some information about the ability of plants to produce their own food. When a child wants to know how socks got covered with burrs while walking in the woods, she needs information about the transportation of seeds.

Photosynthesis

Q: Why are plants always green?

A: There is a green chemical in plants called chlorophyll that allows plants to capture sunlight and mix it with air and water to make food. The food they make is a kind of sugar.

Plants provide the animal world with two things—air to breathe and food to eat. Plants use carbon dioxide, the gas that animals exhale, to produce food. As a by-product of this reaction, they release oxygen, the gas that animals inhale. During the chemical transformation, plants produce the food that all animals must eat to fuel their bodies. That is the difference between plants and animals.

Plants are able to convert light energy to food energy through a process called photosynthesis. They have a special green-colored chemical called chlorophyll in their tissues that accepts energy from sunlight and, in the presence of water and carbon dioxide, convert it into sugar. That sugar can be used immediately by the plant as a source of fuel, or it can be stored in the form of sugar, starch, protein, or oils in plant tissues. These are the plant products that animals must consume to have fuel.

Plants are green because their tissues contain chlorophyll, and any part of a plant that is green in color is carrying out the process of photosynthesis, or at least contains chlorophyll and, therefore, has the capacity to do photosynthesis. Plant tissues that are not green serve some other function besides that of gathering light energy.

What Parents Can Do . . .

Rubbing green leaves on light colored paper makes streaks of green from the chlorophyll pigment. Children enjoy "painting" with leaves. One especially effective, and safe, leaf is spinach. It can be crushed and then rubbed on paper by children as young as two years.

Leafy Structures

Q: Why does my celery have all these strings in it?
A: Celery is the stalk of a plant, just like the stalks on our rhubarb plant. Those strings help the celery hold its leaves up high so that they can get lots of sunlight and air.

Plant parts originally intended by the plant to collect light energy are leafy structures. They're easy to pick out because they are green, and when they're kept whole, they look like leaves. Children often object to eating things like lettuce and spinach by claiming that they don't want to eat leaves—it is very obvious to children what these structures really are, and so far as many children are concerned, they were intended for the private use of rabbits and giraffes.

One of the reasons that leafy structures are so nutritious, though, is that they contain many vitamins and minerals that are actually intended by the plant to be involved in or be produced by the process of photosynthesis. Since these vitamins and minerals cannot be produced in the body of an animal, they must be obtained by eating plant parts or, in some cases, meat from animals that have themselves eaten plant parts.

Many plant parts eaten by people are stems. The plant has a stem so it can get water and minerals from the roots up to the leaves where they can be used with light and carbon dioxide to produce sugar. Those same stems then provide transport of that sugar from the leaves to other parts of the plant where it can be stored. Your child has probably objected to "strings" in celery stalks. Those strings, along with all the water, give the leaf stalk strength and support, and allow it to hold the leaves up in the air where they can get light and carbon dioxide. When the stalk is full of water and strings, it produces a loud snap when it is broken. Of course, if celery is left in the vegetable bin of the refrigerator for too long, it retains its "strings," but loses its "snap." That's because it loses much of its water to the air by evaporation, which is exactly what happens to house plants when they wilt. Their stems lose water that cannot be replaced quickly enough by the roots, and they wilt or "lose their snap." Children are quick to notice these similarities, and they are able to generate their own understanding of what happens with a little help.

What Parents Can Do . . .

Help the child to think of everything he eats that is really a leaf (lettuce, spinach, celery—even parsley). Compare the fact that cows eat mostly leaves, but people like to eat lots of different kinds of food. Human nutritional requirements are quite different from that of any other animals.

Fruits, Vegetables, and Nuts

Q: **Why are the pecans green on the tree and brown on the ground?**
A: They are hiding from the squirrels. Pecan nuts are really seeds, and if a squirrel eats a pecan, the seed will never grow to become a new pecan tree. When they are on the tree, they are green like

the leaves and when they fall to the ground, they are brown like the dirt and old leaves. That makes them hard to see.

A great deal of the fresh produce that is brought home from the store was originally intended by the plant to protect or nourish its seeds. Children are always amazed that hidden inside their apple or orange are tiny seeds, protected by fleshy fruit containing lots of naturally occurring sugar. Children often want to know why apples are so sweet. They are sweet because the sugar produced in the leaves of the apple tree during the warm and sunny spring and summer was stored in the fruit of the apple. But why would an apple tree devote so much energy to making such a beautiful, delicious thing as an apple? Apple trees have been around for a long time; botanists call them one of the "higher plants," meaning they've spent a lot of time becoming very good at what they do; and what they do is make new apple trees. Apples have small brown seeds inside of them, nested inside the hard core, the whole thing surrounded by a firm sweet pulp and then topped with a bright red skin. If that apple fell from the tree and left the seeds to grow directly below its branches, the seeds would have very little chance of developing into a tree. It is too shady, and the parent tree has used up much of the useful materials in the soil in that area already. Apple trees are good parents, just like people, and they wish to start their young ones out on the best foot. Over many centuries, apple trees have developed a means of moving the seeds away to an area that was relatively unexhausted of the things that young apple trees need. They coaxed animals into moving the seeds from them, and they did it by tempting them to eat the sweet fruit. It is easy to spot because it is red, and those tiny seeds, if they are not refused along with the core, pass quite nicely through the animal's digestive tract and are deposited along with a suitable "organic fertilizer."

As anyone with a mulberry tree knows, birds distribute those seeds quite effectively by eating the berries and inadvertently depositing the seeds at a distance from the tree. Many berry seeds are dispersed in this manner.

Pomegranate, pear, and grape seeds are transported via animal intestinal systems. Seeds of oranges, lemons, bananas, pineapple, and watermelon were likewise once distributed in this manner; they originated in a part of the world where monkeys, parrots, and fruit bats lived. These animals were intelligent enough to be able to pull off the skins, allowing the plant to better protect its seeds during their early development while still managing to have them transported.

Young children are also likely to ask other very pertinent questions about fruits, such as "Why are the little grapes in this bunch so sour?" It's because they did not get to grow long enough to get sugar stored in them. A child may also wonder what those little things are all over the outside of her strawberry. They are seeds. Lots of other things fit into this category too—tomatoes, cantalope, pumpkin, yellow squash, peaches, green pepper, avocado, even green beans. Some of them are more obvious than others, but with a little help children can see the seeds in most of them.

There are a few fruits, though, that spend much of their time and energy avoiding being eaten. The seeds they hold inside are very rich in starches and oils and when eaten, are destroyed. Therefore, they don't want to be noticed. These are the nuts, like walnut, acorn, and pecan, that stay green and unobtrusive on the tree and turn brown when they are on the ground. Even when noticed they remain protected, with their hard shells that are often so frustrating to the nut-lover caught without his nutcracker.

What Parents Can Do . . .

Children greatly enjoy searching for the seeds inside fruits, vegetables, and nuts. They learn a great deal when allowed to do this, including such things as the diversity of size and number of seeds within the fruit, whether the seeds are well or poorly protected and the intricate patterns nature has provided tucked away inside many plant structures (stars in apples, circles in bananas, etc).

Roots

Q: Why are there leaves growing out of the top of that carrot?
A: The carrot is really a root and leaves grow out of it. When carrots are in the ground, they store up the sweet food and vitamins that the leaves make up above the ground. That's why we like the way they taste and also why they are good for us.

Roots are not usually green, but then they don't need to be. They are there to absorb water and minerals from the soil, to hold the plant in the ground, and to store sugar or starch for future use. Some roots even have the ability to carry out reproduction, or the production of a new plant. One of these roots is the sweet potato. It is a thickened underground structure that contains large amounts of stored food; that's why people eat them. They are sweet because much of the food stored in them is still in the form of sugar. These fleshy roots can also do something else, though. They can produce new shoots, or tiny plants, when they're subjected to the right conditions.

What Parents Can Do . . .

Allow children to "plant" the top of a carrot and watch the leaves grow back out of the top.

Seeds

Q: Why do you make everything with flour?
A: Flour is ground-up wheat seeds. Wheat seeds are like rice and corn—they are loaded with good proteins, vitamins, and minerals. Plus, flour is good for holding things together when they are cooked.

There is one other type of plant structure that is frequently seen on the table, and that is the seed. Seeds are nothing more than capsules for baby plants. They protect the embryonic plant by holding it securely inside until conditions are just right for its growth and development. Then when the tiny plant does begin to develop, the seed alters its function and becomes a food source for the plant, providing a store of food that is just big enough to last until the young plant can begin to make its own food in its own leaves. The materials that make up the stored food in seeds consist of some of the most important carbohydrates, minerals, and vitamins in people's diets, making seeds one of the most valuable components of a child's meal. Not only that, seeds are meant to retain their nutritional value over a long period of time when kept very dry. Therefore, people can store them for long periods without fear of losing their nutritional value.

It is easy for children to understand the connection between plants and people when they see that fruits and vegetables are truly food for both. It also helps them to grasp the importance of agriculture and gives them a better understanding of the source of their food. Although nutrition is discussed in a different section of this book, it's important to point out here that children are better able to accept the value of a food product beyond its flavor value when they can connect its function in the plant world with their own nutritional needs. All living things have similar nutritional requirements, whether plant or animal. Since children already are so tuned into their bond with other living things, parents can use that bond to sneak in a little information about what is good for them and good for their plant and animal friends.

Seeds come up repeatedly in children's everyday experience. They may notice that right after the lawn is seeded, the birds decide the house is the most popular on the block. They may come in from playing in the vacant lot across the block and notice that their socks are covered with cockleburs. They might notice that those little whirlybirds that fell from the maple tree have a "bean" on the end of them. They may even notice that when they split open their dry roasted peanut, there is a tiny plant inside. All of these items are seeds, and even though they may all look slightly different, they all serve the same basic purpose—they harbor an embryonic plant. Their different structures are due to the different needs of the plant enclosed within the seed. Some contain plants that survive better when they travel far away from the parent plant, and these seeds are equipped with a structure that promotes movement, like the stickers on a burr or the feathery wings of a dandelion seed. Some contain plants that survive best when the seed is planted fairly deep in the soil; these seeds have enough stored food to nourish the plant on the journey to the surface. These seeds are larger, such as lima beans and green peas.

Some seeds are not very picky about being moved or being buried. They are small and light and tend to grow well even when they are just scattered by the wind and on the ground, like grasses and weeds.

There are many different kinds of seeds in a daily diet. Some are obvious, like sunflower seeds and pecans, but others are not so easily recognized as seeds. Cereals are seeds from grass plants like oat, corn, rice, and wheat. Children often find it difficult to classify oatmeal in the same category as their cold breakfast cereals, but it's basically the same thing. It's just not quite as highly processed. Bread is made from flour, and flour is ground–up wheat seeds. The margarine you put on your toast is made of oil pressed from seeds, and many jams come from berries. A few still contain the seeds, although children rarely eat the ones with the seeds in them. Even chocolate comes from a bean—the cocoa bean.

Plants produce the basic foods needed for all living things. As a college professor once said, "If the bugs won't live in it, then there's no point in your eating it either!" All animals look to plant materials to provide all the basic nutritional requirements. Even meat-eaters are consuming muscle which was initially fueled with plant materials. Cows eat grass, and then people eat cows. Earthworms depend on extracting decomposing plant material from soil for their food energy. Pine bark beetles need pine bark; boll weevils need cotton. Sparrows need small seeds. Giraffes spend most of their waking hours reaching for leaves. Plants are the producers in the chain of food-getting. They produce the sugar and starch everything else depends on.

Children notice some of the other things plants do besides providing food. They provide shade, beauty, color. Most of all, in most children's minds, they provide places for animals (including children) to hide. Squirrels and birds live in dens under them. Ladybugs live on grass blades. Toads and frogs live under the low bushes. Spiders set their traps in the holly bushes near the water faucet. Lizards scurry back to the shrubbery when you get out of the car. Butterflies and bees flit from flower to flower, and dragonflies attach themselves to the plants that grow through the water near the lake's edge.

Besides all these things, it does not hurt to remind your child of a few other important things that plants do. Their roots hold the soil, or ground, together and keep it from washing away in a rainstorm. Plants also provide protection from sun and wind. Plants are also just plain pretty. Flowers can capture the attention of children just by virtue of their beautiful colors. Kids see no difference between flowers and weeds; they both have pretty flowers. In fact, weeds would probably win a popularity contest with children—there are more of them around, and no one yells when they pick them!

What Parents Can Do . . .

When you are driving in the country, look at the farmhouses. They are usually bounded by stands of trees that grow in rows on the sides of the house. These trees were not just growing like that naturally; they were planted as wind breaks when the area was first settled.

Flowers

Q: Why do plants have flowers?
A: Flowers are the parts of the plant that make the seed. Seeds are the baby plants that were made by the mother plant inside the flower.

Flowers are beautiful, but that's not the reason they are on the plant. They are there to produce the seeds. Flowers are the reproductive parts of the plant and contain male and female parts just like animals. Pollen, the yellowish powder inside the flowers, is the male cell and it must be brought in contact with the egg, the female cell, for a seed to be produced. The parts responsible for producing these reproductive cells and then for getting them together are in the flower. Bees, birds, and wind serve to carry the pollen to the egg, since plants are rooted to the ground and cannot get their male and female parts together on their own. When they start stirring up all the pollen, many people start sneezing. The eggs are contained in a structure at the base of the flower called the ovary. When the flower dies, it is the ovary that stays behind to be called the fruit, berry, or the pod.

What Parents Can Do . . .

It is great fun to take apart a large flower, such as a sunflower, carnation, or tulip. Children are fascinated by the structures that lie beneath the surface of the flower. It is especially enlightening to observe plants, such as the rosebush, which have blooming and deteriorated flowers side by side. This illustrates the idea that what is left after the flower is a fruit with seeds.

Dangerous Plants

Q: Why do roses have to have thorns?
A: The stems have thorns so that nothing will want to eat them. Thorns protect the plant.

One of the things that often comes up between parents and children is the fact that many plants are unsafe or, at the very least, undesirable to eat. Most good child health and safety manuals have lists of these plants, but none tell parents why these plants are like they are. Children want to know. They remember to be cautious about touching things or tasting things when they are given some sort of explanation for the existence of anything that could possibly want to hurt them. Bees and wasps are viewed with much more respect and less hostility when the child understands their behavior as protective. The same is true for understanding why roses have thorns or poison ivy causes a rash. These kinds of plants have developed a method of keeping animals from eating them. How does a rose know that the child only wants to enjoy its beauty, not its tasty stem? It doesn't, so it presents those nasty thorns to a child as well as to all other animals. Luckily, there are really very few plants with poisonous leaves, berries, or flowers, but since there are some, it's important for children to understand that they should restrict their eating to the things that their parents say are okay to eat. People have lived on earth for a long time, and it was by trial-and-error that our ancestors discerned which plants made people sick when they ate them. They had a few good clues, though—if the animals did not eat them, then they were probably not meant to be eaten. Those plants had something either in them or on them that kept animals from eating them.

What Parents Can Do . . .

Children need to be taught to recognize dangerous plants that exist within their local area. Even three-year-olds can be taught to recognize poison ivy and nettles due to their characteristic leaf arrangements. A plant identification handbook will provide pictures of these and other dangerous plants, useful to both parent and child. And a good rule for children under seven is to never eat berries they find on bushes; they are there for the birds, not children.

Microbiology

Q: What's this stuff floating on your old coffee?
A: That's fungus. It's a colony of tiny living things that landed on the coffee from out of the air. In the air, they are too small to see, but they have grown after eating the coffee, so you can see them.

Children are excellent observers. They are often better at spotting interesting things than adults are. The living things that children observe are usually separated into two groups: plants and animals. Plants are green and do not move, although they grow. Animals grow too and, in addition, they eat and must have the ability to move. But, there is a third category of living organisms that is not so easily observed (although children are quite good at spying the visible ones). These organisms are the protists, the microscopic organisms responsible for decomposing dead organisms as well as for causing disease. The jobs of protists are to decay, to spoil, and to infect. Despite their bad reputation, these tiny organisms are an essential part of all natural cycles.

The members of this group are the fungi, bacteria, algae, and viruses, and often they are so tiny that they can only be seen with a microscope.

Since children have a habit of noticing the least obvious things, they spot these organisms all the time. Taste and smell very rarely inhibit children's inquisitiveness, which is why children must be physically separated from dangerous substances such as lighter fluid and liquid pesticides. That is also why they don't mind looking at things that are rotten.

Fungus or Mold

Q: Why does the bread have cotton on it?
A: That's not cotton; it's bread mold. Look at it with your magnifying glass. There are tiny things like seeds in the air called "spores" that sometimes land on bread. If the bread is kept long enough, the spores have time to grow into colonies of mold. They're usually round and fuzzy, and it's not a good idea to eat them.

It would not be much of a world without fungus. If the green plants of the world are the producers, then fungi are the decomposers. That means they make things rot. All of the nutrients that living organisms need to function are in their bodies when they die, and, luckily, the fungi are around to make sure that those nutrients are given back to the present and future organisms of the world. If nothing ever rotted, the world would be buried in a sea of dead plants and animals. Not only that, the dead ones would have used up all of the nutrients needed to carry on life, so there would be no new organisms to replace them. The job of the fungi is to turn the dead bodies back into the chemicals that made them so future organisms can be nourished.

Molds are like tiny plants. They are part of the larger group of organisms called fungi that also includes mildew, yeasts, and mushrooms. These plantlike organisms are different from green plants because they cannot use light, carbon dioxide, and water to make their own food. They have to eat food by living directly on its surface, and they live on many different food sources. In the home, molds are most likely to be encountered on fruit, bread, clothing, old paper boxes, bathroom fixtures, and even on the body. They manage to get on all these things because they send out tiny microscopic seeds called "spores." Spores are so tiny and light that they simply drift into the air, occasionally bumping into and sticking on a surface that they can eat. When that happens, the spore sprouts a new organism, just like a seed brings forth a new plant. The fungus then forms a whole forest, sending out its root-like filaments, called "mycelia," in a characteristic circular growth pattern. After the fungus colony eats all the food and sends millions of spores into the air, it dies.

Since there are billions of fungal spores in the air, almost everything has spores on it. Even the cleanest homes have them, although defrosting the refrigerator regularly and keeping it cleaned out helps reduce the number in there. Most molds are not harmful, although their spores make many people sneeze when the air gets full of them. Many commercial products are made with fungi, like antibiotics, and even some of the fancy cheeses.

Fungi are fascinating to look at under a magnifying glass. Children enjoy watching the colonies grow. The growth of a fungal colony provides a visual portrayal of the life and death of the fungus community. Children can actually see proof that things go back to their original components (to the child this is usually understood as going back to dirt) when a decomposer, like fungus, breaks down its food source. This decomposition is a wonderful example of a critical cog in the food cycle. Allowing children to inspect a cup of coffee that was left on the workbench for a week will undoubtedly generate much curiosity over the "stuff" on the coffee.

What Parents Can Do . . .

A fun, simple, and educational experiment for children is growing a "bread mold garden." Expose a slice of bread to air for about half a day and then put it in a warm, dark place where it will retain some dampness. Flourishing mold colonies should develop in a few days on that slice of bread. Bread mold, actually green penicillium mold, grows best under dark, warm, and damp conditions. Allow the children to view the fungus with a magnifying glass for extra excitement. (This experiment works best when the bread used is home or bakery baked, not packaged from the grocery store. Many of these brands contain preservatives meant to retard fungal growth.)

Bacteria

Q: Why did you throw away that meat?
A: It had been in the refrigerator too long. That was long enough for tiny organisms called bacteria to grow on it. Since some of those bacteria are not good for you to eat, I decided to throw the meat away instead of taking the chance that bacteria were in it.

Bacteria, like fungi, are decomposers. They feed on other organisms and live in colonies. Unlike fuzzy molds, bacteria colonies look wet and slimy. Bacteria are single-cell organisms that are so small, it would take 500 just to cover the period at the end of this sentence. They generally grow on warm, moist materials derived from living things, and they are most likely to colonize such materials as water, soil, plants, people, and other animals. Bacteria are responsible for causing many serious diseases, but they also serve beneficial purposes such as converting milk to cheese, sour cream, or yogurt.

Viruses

Q: Why has Billy been out of school all week?
A: He is sick with chicken pox. It is a sickness, or disease, caused by a tiny organism called a "virus." Since that virus could make you get the chicken pox, too, Billy is going to stay out of school until his body kills the chicken pox virus. He'll be back next week.

Another infamous protist is the virus. So often parents inform their child that she has to stay home from school because she has a virus without giving her a clue as to what a virus is and what it is doing to make her feel so bad. Viruses are like bacteria, except they are so small that they are not even a whole cell. They have to live in a cell. In fact, no one knew what viruses really were until the electron microscope was invented. It enabled scientists to see incredibly small things, and suddenly it became apparent that some of the mysterious illnesses were being caused by tiny organisms inside the cell.

Algae

Q: What is the green stuff on the side of my aquarium?
A: It's algae. Algae are like tiny plants that all grow together in the water and make a green-colored scum.

Algae is the green scum on an aquarium tank or the long strands of seaweed in the ocean. It is the most like plants of all the protists. Most algae can make their own food; many of them are green, gold, brown, or red (Alabama's Crimson Tide is named for a red algae). They are a very important source of food for many of the animals that live in water. People even use them; one of the substances they produce is used to make ice cream.

Disease-Producing Organisms

Q: Why do you have to wash my cut with that stuff?
A: This is an antiseptic. That means it will clean the germs, or bacteria, out of your cut. Your skin naturally keeps them out, but when it gets cut, the bacteria try to get inside your skin. If they do, the cut gets infected and then it *really* hurts.

Many microscopic organisms, especially bacteria, live inside the human body and serve beneficial purposes. Of course, children rarely hear about those organisms. They are more likely to hear about the ones that cause diseases or infections; these organisms are pathogenic or disease-producing.

All living things are subject to disease, not just man. Most diseases are caused by pathogens, although some others are caused by malnutrition, vitamin deficiency, air pollution, chemical poisoning, genetic defects, and injuries.

Most disease-producing organisms are bacteria, viruses, or fungi. They enter the body of an animal or plant through its openings or through wounds. Skin protects living organisms from being infected by these pathogens, but when organisms are wounded, bacteria, viruses, and fungi that would ordinarily be inactive on the surface of the skin are able to invade the tissues and cells and cause an infection as they grow. Openings into the body, such as the mouth and nose, are common pathways for the entrance of many bacteria and viruses. These openings are also the pathway of transmission, since during sneezes and coughs tiny water droplets are released into the air which may be laden with pathogens.

There are good reasons for insisting that children withstand the discomfort of washing skinned knees and cleaning them with a bit of hydrogen peroxide—these activities attempt to wash away any pathogens that are present and are in a position to enter the body. Even band-aids help in keeping the wound free of enterprising organisms. All of this is also justification for continuing to promote hand washing before meals and nose covering during sneezes. It is also why preschools, day-care centers, and elementary schools insist that children be kept at home when they have fever; fever is a sign of an infection, and its always possible that it can be passed along to others.

Fever

Q: What is a temperature?
A: We say you have a temperature if the inside of you is too warm. It makes you feel bad, but it is your body's way of trying to make you well again.

Fever is actually a defense against infecting organisms. A child's body always has a temperature of about 98°, but when she has a fever, her body temperature is set above the normal. Fever probably speeds up the body's processes that repair damaged cells; at the same time it makes the body an undesirable place for the invading pathogen to live and multiply due to the high heat. Chills are the body making a fever. The muscular action of shivering produces extra heat so the temperature inside the body goes up. The skin remains cool until the chills stop; then the skin warms up quickly as the high temperature radiates out the perimeter of the body. This makes the child perspire.

There are two more defenses in the body's war against pathogens. These are white blood cells and antibodies. Their job is to destroy invading pathogens, or at least make them inactive. A buildup of yellowish pus at the site of a wound is just a sign that a foreign body did indeed sneak in, but the white cells surrounded and destroyed it. That is what pus is—white blood cells. The fact that once a child has had the chicken pox, she can not get it again is proof that her body did exactly what it was supposed to. When it was infected with chicken pox virus the first time, it produced a chemical antibody whose only job was to destroy that particular kind of virus. Once her body has produced the antibody, it will always be there ready to destroy chicken pox virus, even before the disease has a chance to become an infection. The child is now resistant to that particular virus. That is how vaccinations work too. The child is injected with a dead, weak pathogen that signals the body to make antibodies, but does so without causing the disease. Then, since the antibodies are already there, the child will not get the disease.

Children question many of the outward signs left by microorganisms. Therefore, parents should feel free to discuss the existence of the organisms in the interest of satisfying curiosity, despite the fact that the child may have trouble really understanding what she cannot see.

Chemistry

Nutrition

Q: Why do you always make me eat vegetables?

A: Vegetables have vitamins in them. They are there to help the plant grow big and strong, and if you eat them, you will also be able to grow big and strong.

It's often frustrating to read about all of the basic food groups that need to be included in the family's diet. After having read an especially guilt-engendering article in the newspaper about nutrition and health, parents may decide to prepare a meal made of representatives of all four basic food groups, only to have the children complain and go for the peanut butter jar. What good are healthy foods if children won't eat them?

There are many wonderful cookbooks on the bookshelves now that are dedicated to sneaking a good, balanced diet in on children. They do help, but so does the well-informed approach. Children can be persuaded, subversively, that they are knowledgeable ones when it comes to the body and its nutritional needs. When a child asks why he has to drink his milk, he can be challenged; perhaps it has something to do with something white in his body since milk is so white. He'll love to explain—milk is meant to make his teeth whiter. Of course, children do need help to understand that it is actually the calcium in cow's milk going to the teeth and bones to make them harder, but this sly method is often effective.

Think for a moment about the things children wish to avoid. They don't want to get sick and have to stay in bed without T.V. They don't want to break any bones; not that it wouldn't be wonderful to be able to show up at school with an arm in a cast, but it also would mean they would not get up to bat all season in T-ball. They don't want to be smaller than their younger brothers or sisters, and they certainly don't want to have to wear the same size and inevitably have to share clothes with their siblings. They want to be able to see well so they can find the hiders when they play hide-and-seek. They don't want their skin to itch all the time. They don't want to have any teeth removed before cashing them in to the tooth fairy. When given the knowledge, they can search out the vitamins, minerals, carbohydrates, and proteins that will help them to avoid these less than satisfactory situations.

Proteins help make muscles, skin, hair, and nails. Since the human body can't store proteins, people need to eat proteins every day to continue to build these things. Proteins are found in meat, fish, eggs, peanuts, milk, and milk products. Peanut butter and cheese are popular items in this category.

Carbohydrates are very good as quick energy sources for the body. Most of the carbohydrates that children eat are made of sugar or starch, like fruits, candy, bread, and potatoes. Carbohydrates serve as the source of fuel. When there aren't any carbohydrates around, the body will use proteins as fuel sources. This use of proteins needs to be avoided since proteins are better off serving as tissue builders than as fuel to burn. Since the body has a way of storing carbohydrates, they will be stored as fat when they are eaten in excess. People need to balance a diet high in carbohydrates with a lot of exercise. The favorites in this category, not counting the sugary things, are spaghetti, rice, bread, and fruit.

Fat is a word that has gotten a lot of bad press. Fats in the diet are just another energy source. It takes a long time for the body to use up all of the energy in a piece of fat, since fats are broken down in a "slow burn" process. Fats are so energy-rich that they produce twice as much energy as carbohydrates. Foods that have a lot of fat in them are called fatty foods, and people don't have to eat very many of them in order to get all the fat they need. Extra fat in the diet is stored in fatty tissues, and these fatty layers help to protect body organs and insulate the body against heat loss. Therefore, everyone needs some fat in their diet. They're easy to provide because they are in a lot of popular foods like milk, cheese, eggs, butter, and nuts.

Any child who can learn the alphabet can learn about vitamins, since they are very conveniently named after the same. Vitamins do not provide energy; they promote proper body growth and development. The best way to insure that a child will learn to choose a good diet is to help him learn one important thing that each vitamin does and what foods have that vitamin in them. After all, when it comes right down to it, nutrition is a self-serve situation. Regardless of the age, children cannot be force-fed foods that are good for them; they have to want to eat them. They may not like all of them, but there are other reasons to want to eat things than taste preference. Many times children eat one thing to get another (the definition of dinner for many children is what you eat to get dessert). They also eat some things to avoid getting in trouble; the "eat your broccoli or you won't get to go outside after dinner" method. One more method, the one described here (and parents need as many methods as they can find) is the one that encourages the child to drink his milk so that his teeth will be pearly white.

Children can learn that eggs and fish make their stomachs work better and also make them see well by giving them vitamin A. When stomachs work well, they are less likely to get stomachaches, and the better children see things the better they are at playing "seek" games.

Cereals and nuts promote healthy development of the heart and the brain because they have lots of vitamin B in them. Most children don't balk at eating bread or peanut butter, but they still enjoy knowing that they are providing their brains with more thinking power and their hearts with more blood-pumping power. This one is easy to remember, because bread is a "B" word.

All the things that have vitamin C in them, like fresh fruits and juices, help to keep children from getting bruises. They also ensure strong bones and teeth. One of the best ways to get vitamin C is in citrus fruit, appropriately a "C" word, so there really is a reason to drink a little glass of orange juice in the morning.

Vitamin D, the next one down the alphabet, is in eggs, milk, and butter, and most children are willing to eat these things or at least things that are made with them. Vitamin D is similar to vitamin C in that it helps make teeth and bones strong. It is often said that the tooth fairy is apt to leave a little more money for extra strong and white teeth!

The letter K is the next vitamin. It's important because it helps make the blood clot, thereby stopping the blood from coming out when children scratch or cut themselves. Vitamin K is in vegetables and pork, like bacon, ham, and pork chops.

Minerals are also essential to healthy bodies. They are like vitamins because they are not for energy but for good health. Very small amounts of them are needed, but if the body doesn't get a little of each of them every day, it doesn't work quite as well. Calcium and phosphorous, found in milk and grapes, are needed to make strong teeth and bones. Potassium, the mineral in bananas, oranges, and potatoes, keeps muscles working correctly. Iron, found in spinach and watermelon, is needed to make good red blood. These and all the other minerals are supplied, as a rule, by eating any good diet that contains food from the four food categories just described: meat, cereals, fats, and fruits and vegetables.

Children like games, and they like to show how smart they are because it makes parents proud of them. Nutrition can be made fun and, at the same time, children can be started down the road to personal health by letting them tell how nutritious their meals are. An example might be: "How did I do on your dinner? Did I leave any essentials out today? Perhaps I forgot vitamin B or protein." Of course, this method won't always work. There are times when, for weeks, a child will eat nothing but strawberry yogurt, despite all ingenious tricks. It might work, though, and every parent knows that every trick helps.

Water Supply and Waste Disposal

Q: How does this water get to the faucet? Where did it come from?
A: Everyone in our city gets water from a lake just outside of town. It comes to our house through a very large pipe that is underground.

Every living thing on earth needs water. Since new water isn't created, used water must be recycled to fill the water demands of the living things on earth. There is a natural movement of water in the earth's surroundings, purifying it and keeping it coming in an almost steady stream—except for the times when there is too much (a flood) or too little (a drought). Water evaporates, or rises up into the air, from the earth's surface. It stays up in the sky, either invisible or visible (a cloud). It returns to the ground when it condenses (or collects to form raindrops) and falls as rain, snow, sleet, and hail. Living things are welcome to use the water, just as long as they give it back. Even when animals give it back full of sewage, nature accepts it willingly and purifies it. It then can become part of the available water supply again as ground water, rivers, and streams, or precipitation.

Before the age of five, children ask very few questions about water. But after five, they begin to question these cycles with, for example,

"Where does the water come from that comes out of the faucet?" and "Am I drinking the same water that people flush down their toilets?" They are looking for an understanding of people's dependence on their surroundings. They gain assurance when told that it comes from a place near the city where water is either pumped up through water wells or captured from streams and stored in lakes with dams, called reservoirs. From there it is cleaned, or purified, and then sent out in pipes that reach up into each house, bringing in cold water that can be heated by a hot water heater. All of the drains, including the sink, washing machine, shower, and toilet empty into the main sewer pipe under the house. All of the used water from all over the community flows through these sewer pipes to the sewage treatment plant, where it is purified. The treatment plant is efficient at cleaning the water to prepare it to either soak into the ground or evaporate into the air. It is aided by the millions of microscopic organisms that actually delight in munching bits of sewage and waste.

Animals do not live in groups as large as human communities. The amount of waste that animals deposit on the ground is small enough that the resident microscopic organisms in the ground and in the water can chew it up and clean it out. That's their job, and they've been doing it for millions of years. Even human communities avoided the necessity of waste treatment until those communities produced levels of waste that were too extensive for local bacteria to destroy. When water bodies began to generate disease as well as odors, waste treatment was developed. Now the sewage people produce is transformed back into sparkling clean water just as it is for animals. The water cycle is complete.

What Parents Can Do . . .

Children usually have no idea that within the walls of their home are numerous pipes that bring water to the faucets and drain away waste water. Parents can impart a great deal of information by merely pointing out the pipes under the sink, the valve behind the toilet, the clean-out plug on the exterior wall of the home or in the basement, the main water meter (especially when the meter-reader comes by), and the hot water heater.

Especially informative is the plumbing visible in an under-construction home. If one is nearby, children can be walked through and shown the plumbing on a grand scale.

Solids, Liquids, and Gases

Q: What happened to the ice cubes?
A: They melted. The lemonade warmed them up while they cooled the lemonade. But warming them up even a little bit made them turn back into water. Ice is frozen water.

Water conveniently demonstrates the three states of matter: solid, liquid, and gas. Children are quick to notice a change in the state of water—the dog's water may have frozen overnight during the winter; the water level in their play pool may have dropped considerably overnight; the grass may be full of dew in the morning. Although children under seven have difficulty understanding that changes in appearance do not necessarily mean substance changes, they certainly notice the evidence.

All matter takes up space. It exists as solids, liquids, and gases. A physical change in the appearance of matter, such as a change in state, does not change the makeup of its molecules. It merely rearranges them due to a change in energy levels or degree of order. (A chemical change is a change in the makeup of the molecules of a substance resulting in the formation of a new material.)

Solids are easy to see because they have shapes, and their shapes tend to stay the same if left alone. Children understand this characteristic. They are often confused by the wide variety of weights exhibited by solids. Rocks are very heavy, but cardboard boxes are very light. Solids tend to break, although they remain solids even when they are broken into pieces. Some solids, such as ice, for example, can also be easily melted. However, when ice is melted, it is no longer a solid. It is still water, just as ice is, but now it is a liquid.

Liquids are best described to children as things that can be poured, such as water, milk, syrup, hot fudge, and soupy mud. Liquids take up space, just as solids do, but they do not have shapes that remain the same, like solids. The shape of a liquid conforms to the shape of the

container, and when it isn't contained, it spreads out to form a film on a solid surface. Liquids have great play value for young children; they can be poured through funnels and strainers, and they can be mixed to form interesting solutions. During this type of play, children see that liquids exert pressure. They feel the resistance as they attempt to push an inflated ball under the water. They watch closely as some bathtub toys sink below the bubbles while others float. They may even discover that a sibling is more difficult to lift out of the water than when she is in the swimming pool. And knowing that the aquarium must receive extra water occasionally to keep the level consistent demonstrates that evaporation has taken place.

Liquids change into gases when they mix with air. Gases are difficult to discuss with young children due to the fact that they generally cannot be seen. Children do observe the evidence of changes from liquid to gas, such as when water disappears from a chalkboard after it has been wiped. Children generally believe that it has disappeared. They are unlikely to even grasp the idea that air is matter, unless the wind is blowing. There are a few ways of demonstrating the existence of the gas air. An inverted glass can be put into a bowl of water—hardly any water enters the cup. When the cup is tipped slightly, bubbles of air escape showing that there was air in the cup exerting enough pressure on the water to prohibit it from entering the cup. As the air escapes, water can enter. Blowing up a balloon shows that air can take up space. When the air is allowed to escape, it rushes out and can be felt on the face.

What Parents Can Do . . .

Everything in the world is either a gas, a liquid, or a solid. Many objects are composed of all elements of all three states. An apple is a solid, but it very obviously also has liquid in it because it is juicy. It also has a gas component because it smells good when it is eaten. Children can be helped to understand this principle by facilitating their direct contact with physical objects. Questioning or challenging them helps them to describe what they observe and compare it to past experiences.

Chemical Change

Q: What's the orange stuff on this nail?
A: The orange stuff is rust. When the metal in the nail touches air for a long time, the air finally turns some of the metal on the outside of the nail to rust. If you rub the nail on this white cloth, the rust will color the cloth.

The concept of chemical change is very difficult for children under eight to understand. Chemical change results when heat or interaction changes the molecular configuration of a substance, thereby forming a new substance. Evidence of chemical change can be found in rusted iron or steel, tarnished silver, and green copper. In all of these cases, the interaction of substances, over time, has resulted in the formation of a new substance. Chemical change is also evident in burning wood, although in this case heat is the agent responsible for causing the production of a new substance. Children often see the evidence of chemical change and inquire into its development. They are able to grasp the idea that something new was formed as two things came together, but the molecular basis is beyond their understanding at this age due to their inability to conceptualize what cannot be seen (such as molecules).

What Parents Can Do . . .

Allow your children to join in on food preparation, especially cooking. Heating substances and mixing them often produces changes clearly evident to children, such as scrambling eggs, cooking puddings, and whipping cream. Children can also be informed about how soap works or how bleach works. Just by allowing them to join in on daily household tasks, parents can encourage their understanding of the principle of chemical change (as well as many other principles).

Substance Misuse

Q: Can I have some more of the orange medicine? It tastes good.
A: No. Medicine only makes you better when you take the exact amount the doctor said to take. If you take too much, it may make you sick. If you don't take enough, it won't work.

There are many chemical substances evident in children's daily life experience. These substances range from prescription medicine to bubble bath, and even include beverage alcohol and cigarettes. It is extremely important for children to begin to understand at a very early age that many substances are potentially harmful to them and to others if the children eat or drink, or even touch them. In addition, there are also a number of substances that are safe at one level and lethal at another.

The substances that are dangerous for children should always be kept out of reach (discussed in "Accident Proofing"). If a dangerous substance cannot be obtained by the child, she cannot hurt herself with it. However, there are some clues that can be passed on to children age four years and older. Any medicine bottle or capped prescription bottle should be considered dangerous. Children older than three are able to understand the potential risk involved in taking these substances when they understand the meaning of "medicine." They should be made aware of the fact that it is only effective when taken in the prescribed amount. Above that level, any use is actually misuse, and has the potential of causing serious effects. Most medicines are effective in only very small amounts, which makes abuse highly possible.

Another clue to children is the presence of the skull and crossbones, indicating the presence of a highly dangerous poison. This picture should be pointed out to children, and parents should make absolutely sure that the children understand its warning. Children over the age of four are able to recognize the word "CAUTION" which is present on all packages containing ingredients posing potential risks to users. Children who can read may also be told to look for the "KEEP OUT OF REACH OF CHILDREN" warning and the reason it is present. This label is found on such seemingly innocuous products as baby shampoo and furniture polish.

Finally, it is important for children to be aware of the possible harmful effects of beverage alcohol. Children often question why they are barred from drinking these beverages (toddlers are often quite attracted to alcoholic beverages and need to be watched around half-empty beer cans). They deserve to understand the effects of alcohol and the reasons for drinking it. They need to be aware of the potential for abuse. Most of all, they need to understand that the effects of alcohol, as well as most other physiologically active substances, are based on the weight of the user. Children can be told that only adults can drink alcohol and not get sick, because only adults weigh enough. Children don't weigh very much and don't have livers large enough to cleanse the blood of alcohol. Hopefully, an early understanding of the meaning of substance misuse—and abuse—will promote respect for the safe use of chemical substances in later years.

Smoking

Q: Will Uncle Sam's cigarettes make him get sick and die?
A: Cigarette smoking is not good for you, but some people have a very hard time trying not to smoke them. Smoking doesn't make everybody sick; it's just a good idea not to do it at all.

Very young children are now quite aware of the hazards of cigarette smoking. They see public service spots on television, and many even have discussed the subject in preschool or elementary school. The information is given to them to influence future choices about whether or not to smoke. Often, as a consequence, they become afraid that people they love who smoke cigarettes might get very sick or die. Therefore, it is important for parents to address that fear and talk about it with their children. The smoking habit is a personal one, but it is possible that it must be altered to ease a child's fear.

Physics

Q: Why did everyone have to get out of the pool? It's not raining yet.
A: The lifeguard saw some lightning in the sky, and it is dangerous to be touching water when lightning is near. The hot electricity in the lightning will go into everyone who touches the water if lightning strikes the water. Electricity moves easily through water and air, just like electric power moves through electric cords.

When a parent is confronted with a question about the physical laws of nature, he or she is often very uneasy about answering. So many times, the questions asked by little children require answers that easily surpass the child's level of understanding. And, often the parent has difficulty in understanding the principle well enough to answer the child.

Most important is that the parent acknowledge the question. Questions about physical principles signal *excellent* observational skills. The child who asks these kinds of questions has already learned a great deal on her own just by watching. Therefore, any attempt by the parent to enlighten the child will help—the more appropriate and educational the answer, the better. As long as the child is not ignored or doesn't have her question brushed off with a "You wouldn't understand" answer, she will have been rewarded for her efforts. She also will have learned that her parents encourage her efforts to understand the physical principles which govern the world of nature.

Sounds

Q: Why does the pan make a better drum than the cup?
A: It is made of metal, and when you bang on metal, it shakes the air more than when you bang on the plastic cup, so your ears hear it better.

Very young children have the capacity to understand sound. Even children under two can appreciate the sound made by vibrating objects; they can hear sound, as well as see and feel objects that vibrate and produce sound. Early intelligence is related to physical action, and this is a good example of learning through action.

Sound is caused by vibrations. Most homes have an unlimited supply of demonstration materials useful in showing a young child the relationship between sound and vibration. The throat vibrates during speech, and the vibrations feel different during speech or while singing in a high pitch or low pitch. Rubber bands stretched tightly and plucked gently vibrate visibly and produce a musical sound.

Very early in their development, children begin to investigate the nature of sound. They begin to wonder not just what makes sound, but why different things make different sounds, or why some sounds are more easily heard from room to room than others.

Sound travels in waves. Sound waves can travel through many different materials besides air since these waves are caused by forcing

molecules to bump into each other and then pass the energy on. But some kind of medium is required; in space, there is nothing for sound waves to travel in, so there is no sound.

Sounds are not only able to pass through materials other than air, they actually travel more quickly in some—four times faster in water and 15 times faster in steel. Molecules of liquids and solids are closer together than are gas molecules (like air), therefore, it doesn't take as long to pass the energy from one to another. That's why in all the old westerns, the Indian scout put his ear to the ground to get advance warning of the approaching enemy.

What Parents Can Do . . .

Challenge children to make a band, allowing them to use many different types of materials for banging and clanging. Glasses filled with varying volumes of water are especially fun to make music with, although children under the age of four may not be able to participate without causing disaster.

Before the age of three, children fail to connect sound with ears. But by playing whispering games that involve listening carefully or hiding games that require tracking someone by listening to their voice, children begin to understand that ears are for hearing. Especially useful are games that entail covering the young child's ears, first both ears and then one at a time.

Ears

Q: Why do my ears stick out?
A: That way they can capture the sound waves sent out into the air. That's how you hear.

The ear receives sound waves. Most animals have cup-shaped ears that allow them to catch a greater number of sound waves. The shape of animals' ears has a lot to say about their lifestyles. Of course, that understanding will escape a child who doesn't know that ears are for hearing. Asked why a rabbit has long ears, one child replied, "In order to tell it from a ground hog." That was a good answer, considering that he didn't know what ears are used for. After gaining some information, he revised his answer to "they're long so that they'll stick way up above the grass and he can hear the fox coming." His thinking is flawless now that he has a better understanding of the meaning of ears.

Mouths and ears are fascinating to children. Babies stare in amazement when people make funny sounds with their mouths. One-year-olds love to "find Mommy's ear" hidden in her hair. But children do not automatically understand that ears and mouths have anything to do with each other. They know that it is possible to hear sounds, but they often miss the clues that the funny-looking things on the sides of the head are used to pick up those sounds. One friend's four-year-old was convinced that ears were for the expressed purpose of hanging earrings from.

The cup-shaped flaps on the outside of the ear are strategically designed to funnel sound waves down into the ear. They're called auricles, or ear flaps. They funnel sound down into the openings in the ear, which most children think look like dark caves. At the back of each of these "caves" (or ear canals) is the eardrum, a piece of tissue stretched tight like the top of a drum. The sound waves that reach the eardrum cause it to vibrate. The drum then sets into motion a series of bones that make soft sounds louder and loud sounds softer. Finally, the vibrations in the bones are passed to a snail-shaped passage filled with fluid, where tiny hair-like structures respond to the vibrations by sending signals to the brain. It is in the brain where sound is actually perceived.

The eustachian tubes connect the back of the eardrum with the nose and throat, acting as a safety system to keep the eardrum from popping. When pressure outside of the ear is greatly different from the pressure inside, or behind the eardrum, the eardrum gets pushed out or sucked in. By moving air from the outside up through these tubes, the pressure can be equalized, and the eardrum can return to normal. With a change in altitude of air pressure, this equalization causes a funny feeling in the ears. Yawning, swallowing, and chewing gum are movements which help send air up into these tubes, equalizing the pressure. When ears are said to "pop," they are actually saved from popping.

Occasionally, a child may ask why he has two ears. Having two ears allows his brain to tell very slight differences in direction. Children love to play games using this concept. They must have someone hide and then try to find them by listening to sound clues. Even better is playing the game with one and then both ears covered.

Children are amazed to hear their voices on tape. They may not recognize themselves, and they usually wonder why they sound so different. This phenomenon is caused by the brain's ability to hear one's own voice better than anyone else can. The voice is not only heard through the ears, but also through the bones. The skull vibrates during speech and sends the vibration to the fluid behind the eardrum, bypassing the auricle of the ear. Therefore, things are heard in one's own voice that others don't hear. Tape recordings project the voice as others hear it. One thing children love to do that demonstrates this is to bang the tines of a fork against a pan and then hold it to their ears. Then they can do it again and hold it in their teeth. Then try again putting the fork up against other areas, like the forehead, the cheek bone, the breast bone, etc. They are fascinated with the body's ability to hear.

The ear also tells the body it is moving. In the inner ear there is a system of sacks and tubes containing fluid, hairs, and a crystalline dust that together signal the brain of the body's orientation in space. Dizziness is caused by the left-over motion in the inner ear after the body has stopped moving. Young children find great delight in being turned around and around until they can't stand up without tipping over. They often ask why their eyes keep going around when their body has stopped or why they seem to tip sideways for a little while. Those effects are due to the fact that the motion of the inner ear continues for a short time after the body stops moving.

What Parents Can Do . . .

Since children under the age of six or seven (or later) have difficulty imagining something they cannot see, it is fruitless to go into a detailed explanation of the workings of the ear. But keep in mind that mental images of actions stay with them a long time. Let them feel the throat vibrating during a song. Point out differences in animal ears and ask them why they are so different (snakes don't have ear flaps—how do they know when something is approaching?). Point out the ripples on a pond when they drop a rock in, and suggest that sound might be like that. Let them try to play a song with glasses containing different amounts of water. Sing a song and let them listen to it with their ear on your chest (this one might even put them to sleep).

Light

Q: Why does my pencil look broken when I stick it into the aquarium?

A: You see the pencil because your eyes are receiving light that has bounced off the pencil. The light from the pencil in the water moves slower than the light from the part above the water. That makes your eyes see the pencil as if it were broken into two parts. We call that refraction, and you can see the same thing with a straw in a glass of water.

One of the first things to which infants respond is light. Bright lights make them squint, while dim lights allow them to open their eyes and gaze at their surroundings. Moving lights catch their attention. When questions begin to arise, they are first very simple—toddlers may simply say "dark" when they go outside at night or "light" when they discover the power of a light switch. Parents sense, though, in these simple declarations, a developing curiosity.

Children under the age of two limit their interest in the study of light to the presence or absence of it. They notice that they are surrounded by light or by darkness, and their awareness is usually stimulated by a sudden change in conditions—by turning on the light or by stepping outside into the dark at night.

Very young children investigate light by playing with a flashlight, watching a reflection on the wall or by chasing a lightning bug. They may spend a long time turning a wall switch on and off repeatedly. But even though very young children notice light and are able to understand a few elementary things about the nature of light and dark, it is the slightly older child of three to seven who will ask about the actions of light and the occurrence of color.

Children who are beginning to wonder about the nature of light may ask what light really is. They are still too young to be able to completely understand the wave theory of light. Nevertheless, if given a brief explanation supported by plenty of examples as evidence, they can, even at an early age, begin to formulate a picture of the wave theory. This will aid in the discussion of sound as well.

Light is a form of energy. It is produced when electrons, one of the types of particles making up the atom, give up energy. Sometimes the energy is in the form of light, and sometimes it is in other forms such as heat and X-rays. Light energy travels in waves, or rays, and when the eye receives those waves, a picture is seen.

Light waves travel so quickly that light is seen instantly when a lamp or bulb is turned on. Lightning and thunder are useful for demonstrating the speed of light. They occur simultaneously, but the light is seen before the thunder is heard. The light waves travel to the eye faster than the sound waves travel to the ear.

When light waves strike the surface of an object, they may be reflected from the surface, absorbed by the object, or they may pass through the object, making it transparent. The angle at which the light waves strike an object always equals the angle at which they are reflected. Therefore, it is possible to predict where a reflection will be seen, much the same way it can be predicted where a pool ball will travel after it strikes the bumper pad on a pool table.

Refraction is another observable thing that happens to light waves. When they pass from one kind of material into another kind, their speed changes slightly. This causes them to change direction just enough to make them appear to bend. The most obvious example of this is the way a stick or straw appears to bend at the surface when set at an angle in a glass of water. Another puzzler explained by this principle is the way objects in water aren't quite where they appear to be when viewed by looking down into the water—like coins in a fountain or fish in a pond. It's very difficult to explain to a child why it is so hard to catch something that has fallen into the pool. They realize that the object is not exactly where they think it is. One way to show them is to demonstrate with a stick or a straw what effect the water has on the way the object is seen.

Some objects allow light to pass through them but scatter it in all directions, so you can't see through them. These objects are said to be translucent, and children often point out this property in stained glass, frosted glass, and waxed paper.

Other objects allow no light to pass through them; they're called opaque. Opaque objects block the passage of light waves by either reflecting them or absorbing them. Most solid objects are opaque.

Transparent objects with curved surfaces are called lenses. Eyeglasses and magnifying glasses are the most accessible examples of lenses, although they are also used in cameras, telescopes, and microscopes. Lenses bend light in a controlled direction and either make things appear larger (convex) or smaller (concave). Convex lenses are thicker in the middle, and concave lenses are thinner in the middle. Children often notice that a drop of water on a magazine can magnify a printed letter; it is acting as a convex lens.

Children notice light. They sense the difference between light and darkness, but often need some prompting to associate that difference with the sun. It is much easier for them to see that the light in the hallway can afford them an escape from the spooky darkness.

The sun is the most important source of natural light. Sunlight is produced by the extremely high temperatures and pressures in the interior of the sun. The sun is a big ball of continuing nuclear explosions, events very similar to the explosion of a nuclear warhead.

Artificial lights are those light sources which are manufactured and controllable. The most common form of artificial light is the incandescent light bulb, the one that gives off light when a tiny filament, or thread, made of tungsten is supplied with enough energy to make it emit light and heat. Fluorescent lamps are more energy efficient than incandescent because more of the energy used is released as light. Therefore, they are much cooler to the touch. Vapor lamps are used in many windows and signs, and children are fascinated by their cool, soft colors. These lamps have no threads inside of them. They have a gas which emits a specific color when it receives an electrical charge. They give off limited amounts of light, but they are very entertaining to look at.

What Parents Can Do . . .

Children often notice that some animals come out only at night. They wonder why there is a circle of light on the wall, and why it appears to change position as the afternoon wears on. They love to look at diamond rings and comment on their brightness or their ability to make light dance on the wall. They wonder why they can't tell how deep the water is when they can so easily see their diving rings on the bottom, or why it's so hard to grab something that is sinking in the water. They want to know why people wear glasses (more on that later). They can't understand why it is so easy for them to see out of the tinted car windows, but so difficult to see into the car through those same windows. They are fascinated by the fact that they can see a face in the rearview mirror when they're sitting in the backseat. They want to know what happens when a light bulb burns out.

Remember to use what you know to help your children discover the answers to these questions. It's always better for them to find the answer themselves than for you to simply tell them the answer. Hunt down the suspected reflecting object that is causing the circle of light on the wall and investigate its properties. Challenge your child to make a reflection on the wall using another object (you may or may not need to suggest a few possibilities).

Color

Q: How did you make that rainbow with the hose?
A: I put my back to the sun and let its light hit the water. When the light bounces off the water, it comes back to my eyes as a rainbow.

Around the age of two, children begin to learn the names of colors. It's difficult to know if there is any appreciation of color before this age, but once they have learned the names, curiosity about color seems to be set into motion.

Early appreciation of color is just that—appreciation. Most preschoolers could care less about the why's and what's of color. But somewhere around five, they begin to link the colors of their paints and the color of the sky, of rainbows, and of flowers. They notice that on sunny days the pond looks blue and on cloudy days it looks gray. They wonder why lake water looks blue, but when they get a jar of it for their tadpoles, it looks green. They notice that by mixing colors of PlayDoh they can create new colors. They are beginning to inquire into the concept of color.

Rainbows are wonderful teachers. A rainbow can teach the principle of color. It's a beautiful arch of colors that appears in the sky when the sun shines after a rainshower. It can only be seen, though, when the back is to the sun; it only appears in the part of the sky that's opposite the sun. Why is that? Sunlight is white light. It is made up of

all the colors of the spectrum, although they can only be seen when the light is bent, or refracted. Each raindrop acts as a tiny prism, and as the light waves pass from the air into the water of the raindrop, they are bent. Each of the colors of the spectrum are slowed down to a different degree. When the refracted light waves, now separated into all their colors, strike the inner surface of the raindrop, they are reflected back to the eye as the colors of the rainbow. That's why the sun must be behind the back—light must be reflected from the raindrop in order to see the spectrum. A parent can make a rainbow for her child to show him how this principle works by spraying a fine mist of water into the air from the garden hose with her back to the sun. The child may suddenly exclaim, "Look—you're making a rainbow! How did you do that?"

The ROY G. BIV anagram that everyone had to learn in school to memorize the colors of the rainbow is just as workable today. Children enjoy anagrams, and this one moves them a giant step toward understanding color. The colors of the spectrum that are visible to people's eyes are Red, Orange, Yellow, Green, Blue, Indigo (dark blue), and Violet. They appear in this order because of their different tendencies to be bent when entering a prism or any transparent object that refracts light. Red light is bent the least, violet the most, and the others in order in between. One of the most interesting objects that can be given to a child is a prism. She only needs to be shown how to allow light to pass through it. She'll teach herself the rest by playing with it.

Once a child makes a connection between white light and color, she has learned a great deal. Most objects appear gray when viewed in a dark shadow. Colors are also difficult to distinguish at twilight when the sun is almost gone. The reason is that to see the color of an object, light has to be reflected from that object back to the eye. All objects reflect light, absorb light, or allow light to pass through them. Objects that absorb all the light that falls on them are perceived by the eye as black; they reflect no light back to the eye, and, therefore, no color. Red objects absorb all the colors of the spectrum except red, which they reflect. Yellow objects absorb everything but yellow. White objects absorb nothing; they reflect all the colors of white light, and, therefore, they appear white.

Without light, there would be no color. Similarly, without eyes, people could not see color, even if it were there. People depend on their eyes for gathering most of their information about the world around them. That's probably why people have such little ears in comparison with the size of their bodies; seeing is more important than hearing.

A child sees things because light is reflected from an object to her eyes. It's harder to see on a dark, rainy day because there is less light to work with. When afternoon begins to fade into evening, a child may notice that she is having a hard time reading or writing. Twilight sneaks up on her until she realizes that her eyes need more reflected light to work with. When she turns on the light, the colors in the room seem to jump out at her.

What Parents Can Do . . .

Children enjoy the little game of trying to guess the color of an object. This is especially elucidating when played in a dimly lit room. Let the child try to guess the color of objects in the dim light. Then turn on the light and find out if the guesses were right!

Always useful in the investigation of color are water color paints and colored clay, plasticene or PlayDoh. All of these products provide infinite experimentation into color combinations while also providing plenty of fun.

Eyes

Q: Why do I have eyelashes?
A: They help to keep dust and dirt from getting into your eyes.

The eye is like a round bag filled with a clear jelly and surrounded by a tight, whitish coat. There's an opening in the front of this bag called the pupil. Light enters the eye through the pupil; it's the round, black opening in the center of the eye. The iris is the colored part of the eye, and the small muscles of the iris control the amount of light that enters the eye by controlling the size of the pupil. In bright light, the iris closes, making the pupil smaller. In dim light, the iris opens to let more light enter through the pupil.

Children often notice that people have different color eyes. They are constantly receiving comments like "You have blue eyes just like your father" or "Your brown eyes remind me so much of your Grandma."

It's not hard to figure out that your eye color has something to do with the people in your family, but that still doesn't explain why people have colored eyes. The color of the iris has a lot to do with its ability to screen out light. People with brown or black eyes have a better light screen for an iris. In countries nearer the equator, where there is more light, people tend to have dark eyes. Blue-eyed people come from countries where there is less light; their irises don't need to screen out as much.

The front of the eye is covered by a transparent cornea, which acts as a lens, pinpointing light to the center of the eye. Directly behind the iris and the pupil is the real lens. It's a clear disc, about the size of a lima bean, that looks like the lens in a magnifying glass. The lens, along with the cornea, focuses light on a special spot at the back of the eye. This light-sensitive spot is called the retina. It's covered with special cells called rods and cones and also has many nerve endings. When light reaches the rods and cones, they stimulate the nerve endings that send a signal to the brain that you think of as a picture.

Eyes are sensitive and vulnerable objects. For that reason, they sit back in the eye sockets, two bony hollows in the skull. Children can feel the protective bones around the eyes, and it's the shape of this area that has the most to do with what their eyes look like—squinty, dreamy, or beady. The eyes are also protected by hairy structures, the eyelashes and the eyebrows. The eyelid is always ready to shut out intrusions and to help wash the surface of the eye with salty tears. The eye is held in place by six muscles that allow you to move your eyes.

People wear glasses when their own eyes are not doing a good enough job of focusing clear images on the retina. The lenses in eyeglasses refract, or bend, the incoming light just enough to allow the lens to focus an image on the retina. If you put on someone else's glasses, the image on the back of the retina goes out of focus, making it difficult to see correctly.

What Parents Can Do . . .

To help a child imagine something going on inside the eye, have her press lightly on her closed eyelids and then have her tell what she sees. she sees.

The rods and cones are not only sensitive to light, but also pressure. This explains the flashes or twinkles she sees and can be an aid in making an image of what is happening.

Weather

Q: Did the snow grow on the ground while I was asleep?
A: No. It came down from the clouds, just like rain, except that it was so cold outside that the water from the clouds stayed frozen like in the freezer. We got frozen snow instead of rain.

Children wonder about the changing conditions of the atmosphere. Those changes affect even the most basic parts of their daily lives. One day it is beautiful outside, and children are encouraged to turn off the television and go outside to play. The next day it is cold and rainy, and they are made to stay inside. Children wonder how clouds could be puffy and white one day and dark and scary the next. They often assume that thunder and lightning exist for the sole purpose of frightening them. When they wake to find snow on the ground, they wonder if it grew there during the night.

By the age of two years, children have begun to notice the weather. It is often threatening to them, by virtue of its powerful nature. It affects their activities, their dress, their comfort, and even their mood. It is easily observed. Children as young as twelve months old are known to spend a significant percentage of their time gazing out of windows, watching objects move by the force of the wind, observing the way light reflects off many things outside, or just gazing hypnotically at falling rain. Parents recognize their children's interest in weather but often miss the opportunity to capture that curiosity and stimulate it by gently coaxing the child into an understanding of weather.

Children under three years of age certainly cannot yet benefit from hearing about theories of atmospheric pressure and condensation of water (although after three, they can begin to understand even these subjects). When an eighteen-month-old cries "Rain!" and wants his parent to come to the window to share his discovery, he is simply asking for an acknowledgement of his correct observation.

Children between the ages of three and seven are much more astute, though, about the changes in the atmosphere. They are much more likely to ask about the existence of wind and wonder where it comes from. They comment on the fact that white clouds don't rain but gray ones do. They may ask if it is possible to sit on a cloud. These children may be extremely nervous about thunder and lightning and may have imagined terrible things about the meaning of their existence. Parents can expect numerous curiosity-generating situations to arise concerning weather.

Weather is the condition of the air. The entire earth is blanketed by a thick layer of air that's called the atmosphere, but the part of the atmosphere right next to the earth is where the conditions are changeable. This is called the troposphere, and it is where weather occurs. The troposphere is made of water and mostly the gases nitrogen and oxygen. Air is not empty —air is full of air molecules, tiny bits of gaseous materials, cloud droplets and rain, dust, smoke, salt, and ice. Gravity is the force that makes a ball fall toward the earth, and it also causes the number of these air molecules to be highest near the earth's surface. Air pressure is the pressing down of air onto the earth's surface.

Wind

Q: What is wind?

A: It is just air that is moving. Sometimes air moves more than it does other times; then we call it wind. Air moves because it gets crowded in one place, so it moves to a place with a little more room.

Air pressure changes often. Low pressure zones are caused by rising warm air and are often accompanied by clouds and rain or snow. Areas of high pressure indicate areas of descending cool air and usually bring clear skies and cooling spells. Weathermen report air pressure in numbers because it can be measured with a barometer, an instrument that indicates how hard the gases in the air are pressing down on the earth's surface. Air temperature is another thing that changes easily. Air receives its warmth from the surfaces below it. The solar energy from the sun is absorbed by the earth's surfaces, and much of it changes to heat energy and is trapped on earth by the atmosphere. That heat energy warms the air. Water bodies and other moist surface areas keep much of that heat energy, making air above water a bit cooler than the rest of the air. The unequal heating of the earth's surface causes air movement.

Wind is moving air. Air moves from areas of cooler high pressure to areas of warmer low pressure. The speed of the wind depends on the degree of difference in air pressure between the high and the low. It's as if the air molecules in the high pressure area were crowded and decided to move to the low pressure area because it had more room. The more crowded the high pressure areas are, the faster the molecules will move to an available low-pressure area, causing stronger winds. Winds are always named for the direction they come from. North winds come from the North, west winds from the West.

Clouds

Q: How do clouds get in the sky?

A: Clouds are made in the sky when lots of water in the air gets cold and huddles together. So many bits of water get together, or condense, that the sunlight shines off of them and we see a cloud.

The most obvious thing about weather to children is air moisture. Clouds are made of moisture—high, wispy, cirrus clouds are made of ice; middle stratus and cumulus clouds (the ones that look like cotton balls and sheep) are made of water drops. Very low, dark stratus clouds usually mean rain. The amount of water vapor in the air is called humidity. Water vapor is water in the gaseous form, so you can't see it; but you can feel it. Dry air has low humidity—not much water vapor. Moist air has high humidity. As air cools, it becomes less able to hold water vapor, and the gaseous water vapor may condense or form water droplets. The water is now in a liquid state and you can see it—it's called a cloud.

When warm air rises, it begins to get cooled off because it has moved away from the warm earth. When it cools off, the water vapor in it condenses into drops, and clouds begin to form. When enough of the water droplets which form clouds come together, they form water drops so large and heavy that they can no longer stay in the cloud, and it rains. Rain is water from the earth returning to the earth. Water can fall from clouds in other forms, too, such as sleet, snow, and hail. Fog is a cloud that is very close to the ground.

What Parents Can Do . . .

Parents generally know a great deal about the weather, but still struggle with how to explain it to a child. Many children learn this concept with the most ease when it is explained from the vantage point of a single molecule of water, told as if it were a story. Actually, this explanation is a description of the water cycle.

By the age of four years, children are generally able to learn from an imaginary story. This is one about a single molecule of water. It is such a tiny bit of water that no one can see it. It got up into the air when the wind blew across the surface of the lake on a warm day and picked it up. It had lots of extra energy anyway that day from the warm sunlight, and it was ready to jump into the air. People call what it did evaporation. When it got up into the air, there were many other molecules there, not all of them water, but all of them too small for people to see. Despite their invisibility, they were all too crowded together and decided to move to an area where there weren't so many molecules. They all moved so fast that they caused a wind, and people loved it because they could sail their sailboats and fly their kites. When the water molecules finally reached a wide open airspace, they were pretty light, and they started to rise up into the sky like balloons. As they moved way up into the air, it got cooler because the warm ground was being left far below. They became so cool that they started searching for other companions, other water molecules, to huddle with around a tiny particle of dust. Pretty soon there were so many of them stuck together that they formed a droplet, a gathering of water molecules big enough to see. People call this "condensation." Now that they are together in a droplet way up in the sky, they freeze into an ice crystal. Other molecules of water can now freeze onto them until there are enough of them together to look like a cloud. Eventually, they begin to fall as a snowflake. They start to clump together with other snowflakes as they get heavier and fall down through the air. If they fall into warm air, they melt and become rain. If the air they fall through stays cold all the way to the ground, they stay as a snowflake and become

snow. No one is exactly sure what happens to them to make hail, but it probably has something to do with being thrown back up into the cloud over and over until they become a ball of ice.

This little story has helped many children understand weather. It also allows them to project their feelings onto the elements of weather, something they're bound and determined to do in their own attempts to understand. It can help explain dew, since it is the cooling of surfaces to the point where water molecules will gather on it and form droplets. This happens on grass, leaves, car windshields, and spider webs after a cool night. It can explain fog on bathroom mirrors, since warm water in the tub gives enough energy to many molecules that they escape into the air, only to become droplets again when they are chilled on contact with the cool mirror. It is possible to see how one puzzle piece can complete many pictures. One bit of understanding provided by a parent can help a child to understand many other puzzling situations.

Lightning and Thunder

Q: Why do we have to have thunder?

A: Thunder is just the sound we hear when there is lightning. Lightning happens when the clouds make a big light flash, just like a lightbulb. The lightning flash is so hot, it kind of makes the air boom out fast enough to make a sound like an explosion. Really, it's just very fast-moving air that makes the boom. Thunder cannot hurt you.

The first two things children notice about weather are thunder and lightning. They generate such different feelings than do rain, wind, and snow that it hardly seems appropriate to put them in the section about weather. Most children are fascinated by changes in the weather, but are fearful of the thunder and lightning. There are no guarantees that by gaining an understanding of lightning and thunder, children will cease to be afraid of them. Many parents can't seem to shake that funny feeling in their stomach when they're in the middle of an electric storm, but it is important that they try not to pass that fear on to their children. Armed with the facts about what lightning and thunder are and how they came to be there, parents should be able to calm their children's anxiety and at the same time teach them a little more about weather.

Unequal heating of the earth's surface causes air to be heated unequally. Areas of warmed air rise in columns until they reach an altitude where the temperature is cool enough to chill the water vapor in the air and cause it to condense. That's how the clouds (in this case, thunderheads) are formed. The air that is rising in the thunderheads is moving at a very high speed and is so strong that, when raindrops begin to form, some of them are unable to travel down to the ground against the current of the rising air. As these drops are forced upwards, many of them are torn apart. The larger droplets continue to fall down while the smaller droplets are carried up. When the raindrops are torn apart, the larger droplets become charged with positive electricity while the small ones are charged negatively. Soon the entire cloud is electrically charged, with some areas charged positively and some charged negatively. This is the same sort of situation that takes place in an electric current. But in a current, electrical charges have wires to move through and, therefore, the power can be concentrated in one place, like in a light bulb or a toaster. Electrical charges in thunderheads have no wires to move through. In fact, air is a good insulator, meaning it resists the movement of electrical charges. So the positive and negative charges in the clouds just become larger and larger until they finally overcome the ability of the air to resist their movement. A stream of charged air suddenly begins to flow from one area of the cloud to another or from the cloud to the ground, and the electrical charge begins to flow along with it. This great surge of current through air is called "lightning." The temperature of a lightning flash is believed to be about as hot as the surface of the sun. The sudden, tremendous heat and corresponding rapid expansion of air are so great that your ears hear a crash and a rumble. Since light travels faster than sound, you see the lightning before you hear the thunder, although they happen at the same time. Most people probably learned as a child how to count the seconds between the flash and the thunder in order to measure how far away you are from the flash. If not, remember that the flash was one mile away for every five seconds that elapsed between the flash and the thunder. Really close lightning causes thunder that is loud and sharp.

Long rumbles are caused by echoes and by the long, crooked path of lightning, since the thunder of a long lightning bolt doesn't all get to the ear at the same time.

Children, especially younger than six, have great difficulty understanding the theory of electricity. Although it's important for parents to have an understanding of lightning based on the presence of positive and negative charges, their children will feel no better about storms after a lecture on electrical charges. Children do understand the power of electricity, though. It's easy to see in bulbs, in electrical appliances, and in instances of static electricity, like when their hair is attracted to the comb or when their blankets crackle when they get in bed at night. One of those examples might be used to try to show the power of an electrical circuit. A parent could turn on the toaster and show the child how the wires light up, making sure he sees the wires before, during, and after heating, then explain that strong rain storms sometimes make the cloud act like the electricity in the toaster, but there are no wires to use as pathways. The electricity in the clouds has to make its own path, and it is seen as a bolt of lightning or a flash of light. There is heat next to the light just like in the toaster, only there is so much heat that the air moves away very, very fast—so fast that it sounds like an explosion called thunder. It's all a message from the clouds that it's going to rain. Tell the children also, for safety's sake, that lightning could touch them, and it is so powerful that it is dangerous to people—like the toaster or the electrical outlet, only worse. As long as they are where it won't touch them, though, like inside the house or car, under a ledge or if nothing else is available, even in a ditch, it is really not dangerous. It may be bright and loud, but it won't hurt them if they don't touch it. Hopefully, they'll feel better when they understand; maybe they'll even go back to sleep!

Heat

Q: How did the car seat get so hot?
A: The sun was shining on it. The sun not only gives us light, it gives us heat too. When the sun shines on something, that thing it is shining on soaks up heat from the sun.

Children often want to know what makes hot things hot. The answer is sometimes easy, such as an object is next to a flame. Without really understanding the physics of heat, children can appreciate the fact that fire makes things hot. It's not so easy to explain why the palms of their hands get hot when they rub them together, or why the seat belt becomes too hot to touch after an hour in the sun. Heat is not a difficult concept, but many times requires a bit of creative thinking to get the explanation down to a child's level of understanding.

Heat is the energy of motion. When the palms of the hands are rubbed together, the molecules in the surface of the skin bump into each other and are made to move. With continued rubbing, the molecules move faster and faster and the energy that is expended is given off as heat. Everything that gets hot—the vinyl in car seats, the toaster, the sidewalk, the light bulb—has molecules that are moving fast. Heat, then, is the energy of moving molecules.

The theory of heat is extremely difficult for a child. What on earth is a molecule? The molecular theory is the idea that there are tiny bits of water which, when grouped together in large numbers, make the substance people recognize as water. When one single individual particle is divided, it is no longer water; it has become two new substances. A "molecule" is the limit that is reached when a substance is divided right down to its very last bit. If it is broken apart beyond that point, it is no longer a piece of the substance. It is a different molecule.

The molecular theory is one of those things that children can talk about as if they understand when actually they can only barely grasp. A child may nod her head in understanding when she is told that all things are made up of tiny particles of a substance, called molecules. She doubts if anyone would lie to her, but she finds it difficult to believe the idea that what looks continuous to her, such as a glass of water, is actually made up of exceedingly smaller bits of water, right down to a single bit of water. There are a few ways to demonstrate this idea. Take her out to see the sidewalk and ask her what it is made of. Then get a magnifying glass and look at it again. Lo and behold, it's made of tiny bits of rock or sand. There are particles in the cement that can't be seen until they are examined closely. Another method to show the possible existence of molecules is to put a sugar cube into a glass of warm water. Take a sip and then take another one when the sugar cube has dissolved. The sugar in the water can be tasted, but the cube can no longer be seen. That's because the molecules of sugar flew off the sugar cube into the water. The warm water gave the molecules extra energy, enough to allow them to jump off the cube.

Children over four years of age begin to grasp some measure of this theory, although they may lack the understanding that would allow them to apply the theory in another situation. They enjoy acting out the energy of molecules by jumping, running, and dancing. Energy makes everyone want to get up and move around. The same principle applies to molecules. Heat energy makes them move around and take up more space. When someone has a fever, it means they have so much extra heat in their mouth or under their arms that they make the mercury molecules in the thermometer move around and take up more space. The only space available is up the tube, so the line goes up as mercury moves up the tube. It's the same principle when the skin warms near the campfire. Heat moves because it keeps warming up adjacent molecules, setting them into motion, and causing them to release heat. Most of a child's questions about heat can be answered with these bits of information.

What Parents Can Do . . .

One graphic illustration of the energy heat gives to molecules is also very fun and nutritious. When popcorn is allowed to pop without the lid on the popper, it serves as a demonstration of two molecules getting the energy to leave a surface (such as the sugar cube experiment). To do this demonstration and avoid burns from the hot oil, put the corn popper on a large surface (like the kitchen table) that has been covered with newspaper. Put more than the usual amount of corn in the popper (but the normal amount of oil). Begin the popping process with the lid on, but when the popped corn begins to push on the lid, remove it and let the excess corn "pop" out and escape. Then eat the experiment!

Electricity

Q: Why can't I plug in the record player?
A: Until you're five years old, I don't think it is safe for you to plug things in. It's difficult to make the plug fit in the socket, and if you don't get it in correctly, or if you accidently put something in the socket that is not a plug, you could be hurt badly. Electricity comes from those sockets; it makes many machines work, but it also is strong enough to burn people.

Until the age of three years, it is enough for a child to know that the outlet is hot and that if she touches it her parents will be upset. It is impossible to explain to a child under the age of three anything about the meaning of electricity. Parents' efforts are better directed at keeping their child from injuring herself. There is no way to be certain that a toddler will not put a nail into an electrical socket, even when his parent has made an earnest effort to show him that only plugs go into sockets.

Young children begin, around the age of three or four, to suspect that there is a great deal of power in electrical outlets and switches. It may occur to them that the outlet is something other than hot if it will make a favorite program appear on the television screen.

They may begin to attempt to manipulate these dangerous objects again, although, in this case, the motivation is different. They want to know how to use them. It has become obvious to them that electrical outlets have the power to make things work. Things around the house that are important to them must be plugged into the outlets or they do not work, like the night light, the record player, and the family television. Anything that is so powerful is bound to set the mind in motion.

Now, they are ready for an explanation of electricity—what it is, where it comes from, and how it does its work.

Since electricity is most easily understood using the model of the atom, that method of description will be used here. Remember, though, that children under the age of eleven and twelve are not able to or have great difficulty in imagining something that they cannot see, so it is difficult, often useless, before that age to try to explain electricity to them using the model of an atom. There are other ways of getting some understanding to them. It is first helpful to refresh the memory of the electrical principle.

The nucleus of an atom contains many positive electric charges called "protons." Around that positively charged nucleus is a cloud of negative charges called "electrons." When these electrons move from one atom to another, it is called "electricity." An electric current is the flow of these electrons through some kind of material. Of course, free electrons do not pass through materials smoothly. They bump into the atoms of the wire or other material and their progress is slowed down. This braking effect is known as resistance. The lower the resistance, the easier the flow of electrical current. A conductor is a material that allows electricity to pass through it easily, such as the wires within an electric cord. Materials that do not allow electricity to flow through them are called insulators, like the plastic covering around the electric cord.

The resistance effect is responsible for the light in a bulb as well as the heat in the toaster. Electrons do not freely pass through materials, they bounce off the atoms of that material. Thin wires produce high resistance since there is a very limited area of wire for the electrons to pass through. Constant collisions cause the atoms within the wire to vibrate, giving off heat and light.

There is another kind of electricity besides electrical currents; it's called "static electricity." It occurs when an object has an unequal number of protons and electrons. It is often experienced while doing the laundry.

Some of the clothes in the dryer have extra positive charges and some have extra negative charges. Since objects with opposite electrical charges attract each other, the positively charged sock sticks to the negatively charged nightie.

A "circuit" is a pathway along which electrons can flow. Electrons won't flow, though, unless they can flow from a place where there are extra electrons (a negative charge) to a place where there aren't enough electrons (a positive charge). If their path is blocked along the way, they won't move. Electrical outlets are a break in the pathway of electricity. The power that comes into the home allows electrons to move from a negatively charged place to a positively charged place as soon as something is plugged into the outlet that completes the pathway, or circuit.

Most items that are plugged in at home have a switch. It's a device that completes or interrupts the circuit by switching it on or off. The power company measures how much electricity is allowed to flow through the home by putting a meter outside. The electric meter measures the amount of electricity, in kilowatts, that are used per hour. Children love to watch the meter and are able to grasp some measure of understanding when a parent explains a bit about electricity and shows them the meter.

Unfortunately, the human body is a good conductor of electricity, which means that electric currents flow easily through it. The electrical outlet allows the electric current to flow through a bulb and make it light (show this to the child by plugging in a light which has already been switched "on"). Children need to understand that light bulbs, televisions, and record players don't have nerves, so electricity doesn't hurt them. People do have nerves, and the nerves serve as the body's messenger service telling each part of the body to do what it should. The messages they carry are tiny electrical charges, so in a way, they are like little wires. The problem with household electricity is that the amount that comes into the house through the electrical outlets is a great deal more than the body uses. It is so much more that when it accidentally flows through the body, it causes shocks which dangerously, even fatally, disturb the body's nerves so that they can't do their jobs, like making the heart beat or making the lungs breathe. It also causes damage to many of the tissues in the body. A child should be made to understand that under no circumstances should she ever put herself in the position of possibly allowing that electrical current to flow through her body; it is meant for things with cords, not for things with feelings.

What Parents Can Do . . .

Here is a list of some good safety rules to give children. Children over four can generally be trusted to follow them when they understand the reasons behind the rules.

1. Never put anything into an electrical outlet that doesn't have an electric plug.
2. Do not touch an object that uses electricity when you are wet, or even touch the switch that turns it on and off.

3. Do not touch an electric wire or power line that has come off the pole or is laying on the ground.
4. Only experiment with electrical systems when an adult can help you.
5. Do not play out in the open when there is lightning.
6. Never fly kites or climb trees near electric power lines.
7. Never put any part of the electric device into your mouth!

Simple Machines

Q: Why do the moving men slide everything down that board?

A: That board is a ramp, a kind of inclined plane. It is much easier for them to slide the boxes down that ramp than it is for them to lift them off of the truck.

The average home is full of simple machines. Hammers, nuts and bolts, scissors, and brooms are all simple machines. Children can be helped to understand important scientific principles by encouraging their investigations into such items. Many toys are even machines, or they employ one or more machines as a working part—slides are inclined planes, see-saws are a type of lever, bicycles employ a gear and chain.

What Parents Can Do . . .

Children easily accept the idea that their toys slide down inclined planes; this principle can even be extended to include the idea that it involves less work to push a heavy load up an inclined plane than it does to lift it. Levers and pulleys are sources of great enjoyment for young children as they learn to lift different objects using these simple machines. These experiences build up mental images which can later be drawn upon in a problem-solving situation, one of the keystones of scientific investigation.

Space

Q: Why does my ball always come back down when I throw it up in the air?

A: There is a rule, a law of nature, which states that everything tossed or propelled up in the air will come back down to earth.

This particular law is called "gravity." Laws of nature have existed since the beginning of time and govern all physical phenomena of our world and the universe beyond. One of the jobs of scientists is to understand and apply the laws we do know about and to discover new laws. Over the centuries we have come to understand many of these laws, but there are still more laws yet to be discovered and understood. Most of the laws are expressed in equations and these equations can be proved and worked out over and over with the same result. This means that the equations can be used to predict events in nature like the coming of a comet or that a ball will always fall back to earth. Traffic laws are expressed in words and nature's laws are expressed in equations.

Children are great believers in rules since their lives are governed by them. Moral rules allow them to predict whether or not an action can be considered right or wrong. It is *very* natural, therefore, for children to perceive that nature and the inanimate world has rules, also.

Although the theory governing the law of nature in question may be beyond the child's ability to understand, it is important for children to know that there is an exciting story behind the discovery of the law and that as the child's experiences with nature multiply, the theories will reveal themselves with ease. At first however, children are generally satisfied to at least know there is a rule and the opportunity should not be lost to emphasize the connection between these rules and the rules of mathematics—equations.

Q: How do space shuttles get into space?
A: A space ship must be propelled upward, against the force of gravity, past the earth's atmosphere before it can move into space. The rockets that push space shuttles upward have to burn fuel very fast and very hot to get up the power for the big push. This power is called thrust and it has the force of an exploding bomb but the force is channeled and controlled. It is not unlike the effect of letting air out of a toy balloon and the balloon flies away.

Space is a frontier and, like all frontiers, it is dangerous. It is not a very friendly place and it is the job of scientists to learn how to get around the laws of nature to make space a safer place for exploration. This is what makes so much of science daring and exciting. People with inquisitive minds will always be willing to take a risk in an attempt to discover more about the laws of nature.

Q: What is a star?

A: A star is a giant ball of gas, like our sun, burning at tremendously high temperatures. The star is so hot that we can see the glow from its burning gas millions of miles away. The energy for the burning gas is deep inside the star from enormous gravity pressures pushing in on its core.

Stars are really nuclear reactions. In the star core, billions and billions of tiny particles bump against each other and melt together and then break apart and disappear. The starlight we see is from this fusing and splitting that causes a continuous reaction held in place by the star's massive gravity.

Shooting stars, or meteorites, are not really stars. They are small pieces of rock or metal that burn up when they hit the air upon entering the earth's atmosphere. They burn with enough intensity to generate visible light.

Planets do not burn and are not like the stars. We see only reflected light of the sun on our planets in the solar system.

Children see stars as tiny specks of light in the night and early morning skies. They can be helped to understand the true nature of stars by comparing them to the sun. Stars cannot be seen in the day because the sun is shining just as it's difficult to tell if a lamp is on in the day if the room is naturally bright with sunlight. They can be shown that just as the sun makes things warm, the light from stars is light from heat.

Inadvertent Lessons: reading to your child

It has been my personal experience, as a parent as well as a teacher, that children learn a great deal about science from the stories that adults read aloud to them. I don't mean the "Let's Find Out About Machines" type series of science inquiry books. I mean the storybooks in which something happens to people or animals or things—something interesting, exciting, or emotional. It's stories like these that can hold even the most "fidgety" child's mental attention and can print an indelible lesson on his memory banks. The educational value of these storybooks is unimportant to the child—he just wants to know what happens to the characters. But often a curiosity-generating situation stumbled into during the day can be reinforced by a cleverly chosen storybook at bedtime. For this reason, I have listed a few storybooks that I consider highly appealing, appropriate, and educational for children from toddler through about six or seven years. You'll find that these books casually express important scientific theories that have been touched upon in this book. Their lessons are cleverly disguised in intriguing stories of animals and people—you and your children will have fun finding those hidden theories and applying them to your daily life experiences.

- **Animals**

Amos and Boris by William Steig. Farrar, Straus and Giroux, 1971.
Animals Born Alive and Well by Ruth Heller. Grosset and Dunlap, 1982.

Blueberries for Sal by Robert McCloskey. Viking, 1948.
Spiders in the Fruit Cellar by Barbara M. Joosse. Alfred A. Knopf, Inc.
The Very Busy Spider by Eric Carle. Philomel Books, 1984.
The Very Hungry Caterpillar by Eric Carle. Collins Publishers, Inc., 1979.
Wheel on the Chimney by Margaret Wise Brown. Harper and Row, 1948.
Your Owl Friend by Cresent Dragonwagon. Harper and Row, 1977.

• Animals and Seasons

All Year Long by Nancy Tafuri. Penguine Books, 1983.
Big John Turtle by Russell Hoban. Holt, Rinehart and Wilson, 1983.
First Comes Spring by Anne Rockwell. Thomas Y. Crowell, 1985.
Frederick by Leo Lionni. Pantheon, 1967.
The Big Snow by Berta and Elmer Hader. MacMillan Company, 1948.
The Year at Maple Hill Farm by Alice and Martin Provensen. A. Hoen and Company, 1978.
Winter Magic by Eveline Hasler. William Morrow and Co., 1984.

• Death

Badger's Parting Gifts by Susan Varley. Lothrop, Lee and Shepard Books of William Morrow and Company, 1984.

• Sun and Moon

Grandfather Twilight by Barbara Berger. Philomel Books, 1984.
Many Moons by James Thurber. Harcourt, Brace and World, 1943.
Nothing Sticks Like a Shadow by Ann Tompert. Houghten Mifflin Company, 1984.
The Nightime Book by Mauri Kunnas. Crown Publishing, 1984.
Wait Til the Moon is Full by Margaret Wise Brown. Harper and Row, 1948.

• Light and Color

A Firefly Named Torchy by Bernard Waber. Houghton Mifflin Company, 1970.
Dawn by Uri Shulevitz. McGraw Hill Ryerson Limited, 1974.
The Rain Puddle by Adelaide Hall. Lothrop, Lee and Shepard Company, Incorporated, 1965.

- **Weather**

Flash, Crash, Rumble and Roll by Franklyn M. Branley. Harper and Row, 1985.
The Snowy Day by Ezra Jack Keats. Viking Press, 1962.

- **Plants**

A Tree is Nice by Janice May Undry. Harper and Row, 1956.
The Carrot Seed by Ruth Krauss. Harper and Row, 1945.
How My Garden Grew by Anne and Harlow Rockwell. MacMillan, 1982.
The Rose In My Garden by Arnold Lobel. Greenwillow Books, 1984.
This Year's Garden by Cynthia Rylant. Bradbury Press, 1984.

- **Sound**

Secret Sounds Around the House by Marthe Seguin-Fontes. Larousse Company, 1983.

- **Simple Machines**

Locks and Keys by Gail Gibbons. Thomas Y. Crowell, 1980.
Machines by Anne and Harlow Rockwell. MacMillan Publishing Company, Inc., 1972.

- **Anatomy**

Marvelous Me. All About the Human Body by Dr. Anne Townsend. Leon Publishing, 1984.

- **Teeth**

The Berenstein Bears Visit the Dentist by Stan and Jan Berenstein. Random House, 1981.
How Many Teeth? by Paul Showers. Thomas Y. Crowell, 1962.

Bibliography

Allison, Linda. *Blood and Guts*. Boston: Little, Brown and Company, 1976.

Allison, Linda. *The Reasons for Seasons*. Boston: Little, Brown and Company, 1975.

Allison, Linda. *The Wild Inside*. New York: Sierra Club Books, Charles Scribner's Sons, 1979.

Asimov, Isaac. *Please Explain*. Boston: Houghton Mifflin Company, 1973.

Besser, Marianne. *Growing up with Science*. New York: McGraw-Hill Book Company, Inc., 1960.

Blough, Glenn O. *You and Your Child and Science*. Washington, D.C.: Department of Elementary School Principals, National Science Teachers Association, 1963.

Blough, Glenn and Julius Schwartz. *Elementary School Science and How to Teach It*. New York: Henry Holt and Company, 1958.

Brown, Sam Ed. *Bubbles, Rainbows and Worms*. Mt. Ranier, Maryland: Gryphon House, Incorporated, 1981.

Caplan, Frank. *The First Twelve Months of Life*. Toronto: Bantam Books, 1971.

Carin, Arthur A. and Robert B. Sund. *Developing Questioning Techniques*. Columbus, Ohio: Charles E. Merrill Publishing Company, 1971.

Carin, Arthur A. and Robert B. Sund. *Teaching Science Through Discovery*. Columbus, Ohio: Charles E. Merrill Books, Inc., 1974.

Devoe, Alan. *This Fascinating Animal World*. New York: McGraw-Hill Book Company, 1951.

Formanek, Ruth and Anita Gurian. *Why? Children's Questions. What They Mean and How to Answer Them*. Boston: Houghten Mifflen Co., 1980.

Gesell, Arnold, Henry Halverson, Frances Ilg, Helen Thompson, Burton Castner, Louise Bales Ames, and Catherine Amatruda. *The First Five Years of Life*. New York and London: Harper and Brothers Publishing, 1940.

Gesell, Arnold and France L. Slg. *The Child from Five to Ten*. London: Hamesh Hamilton Ltd., 1946.

Good, Ronald G. *How Children Learn Science*. New York: MacMillan Publishing Co., Inc., 1977.

Lonetto, Richard. *Children's Conceptions of Death*. New York: Springer Publishing Company, 1980.

McCandeless, Boyd R. *Children: Behavior and Development*. New York: Holt, Rinehart and Winston, Inc., 1967.

A Nation at Risk: The Imperative for Educational Reform. National Commission on Excellence in Education. U. S. Government Printing Office. Washington, DC, April 1983.

"An N.S.T.A. Position Statement. Science-Technology-Society: Science Education for the 80s." *Science and Children*, 22, 4, 1985.

Nickelsburg, Janet. *Nature Activities for Early Childhood*. Menlo Park, California: Addison-Wesley Publishing Company, 1976.

Piaget, Jean. *The Child's Conception of the World*. London: Routledge and Kegan Paul Limited, 1929.

Postman, Neil and Charles Weingartner. *Teaching as a Subversive Activity*. New York: Delacorte Press, 1969.

Saunders, Ruth and Ann M. Bingham-Newman. *Piagetian Perspectives for Preschoolers*. Englewood Cliffs, New Jersey: Prentice Hall Inc., 1984.

Trelease, Jim. *The Read-Aloud Handbook*. New York: Penguin Books, 1982.

Trojack, Doris A. *Science With Children*. New York: McGraw-Hill Book Company, 1979.

Troneck, Edward and Lauren Anderson. *Babies as People*. New York: MacMillan Publishing Company, 1980.

Wadsworth, Barry. *Piaget for the Classroom Teacher*. New York: Longman, Incorporated, 1978.

White, Burton. *The First Three Years of Life*. Englewood Cliffs, New Jersey: Prentice-Hall, 1975.

Wolfinger, D. M. "The effect of science teaching on the young child's concept of Piagetian physical causality: Animism and Dynamism." *Journal of Research in Science Teaching*, 19, 7, 1982.

Wolfinger, Donna M., *Teaching Science in the Elementary School—Content, Process and Attitude*. Boston and Toronto, Little Brown & Co., 1984.

Yager, Robert E. "The Major Crises in Science Education." *School Science and Mathematics*. 84, 3, 1984.

Index

Accident proofing, 21–24
Air pressure, 141–142
Algae, 113
Animals, 27
   behavior of, 82–83
   body coverings, 73–77
   cold-blooded, 96
   communities of, 88–89
   dangerous, 90–92
   defined, 72
   parasitic, 89–90
   seasons and, 94–97
   temporary care of, 92–94
   warm-blooded, 96–97
   See also Zoology
Animism, 29, 44
Antibodies, 115
Antiseptics, 113–114
Apples, 102
Appliances, 22

Arithmetic, 10–11
Autumn, 96

Bacteria, 112
Balancing skills, 32
Behavior
   animal, 82–83
   imitative, 21
   inborn, 82–83
   learned, 83–84
   negative, 32
   social, 21
Biology, 45–71
Birds, 75
Blood, 52–55
Body
   coverings for, 73–77
      camouflage, 77–81
      feathers, 74–75
      fur, 75

scales, 74
parts of, 81–82
temperature of, 114–115
Bones, 49
Botany, 97–108. *See also* Plants
Bowel movements, 61
Brains, 63–64
Bread mold, 109
Breathing, 55–57

Camouflage, 77–81
Capillaries, 54
Carbohydrates, 117
Carbon dioxide, 56–57
Carnivores, 85
Carrots, 103
Celery, 99
Chemical change, 124
Chemistry, 115–126
Chlorophyll, 99
Cigarette smoking, 126
Circuits, 151–152
Classification skills, 20, 28, 40, 43
Clouds, 142–145
Cold-blooded animals, 96
Color, 135–138
Coloration, animal, 80–81
Communication skills, 40
Comparison skills, 34, 36
Construction, 11
Cornea, 48
Counting activities, 10–11, 24
Creation, 70–71

Dangerous animals, 90–92
Dangerous plants, 108
Death, 39, 66–70
Decay, 87–88
Deciduous teeth, 61
Diet, 115–119
Digestion, 58–61

Disease-producing organisms, 113–114
Dynamism, 29–30

Ears, 129–131
Ecology, 84–87
Electrical safety, 21–22, 126–127, 152–153
Electricity, 151–153
Electrons, 151
Eustachian tubes, 130
Exploratory play, 26
Eyes, 138–140

Fall, 96
Fats, 117
Fears, 38–39
Feathers, 74–75
Fever, 114–115
Fingerprints, 47
Fish, 78–79
Flour, 104
Flowers, 107
Food, 58–61
Food chain, 85, 87
Fruits, 100–103
Fungus, 109–112
Fur, 75

Games, 19, 24
Gases, 121–123
Gender, 39–40, 65–66
Grazers, 85

Hair, 47–49
Heat, 147–149
Helpful relationships, 90
Herbivores, 85
Houseplants, 22
Human biology, 45–71

Idea formation, 37–44
Imitative behavior, 21

Inborn behavior, 82–83
Infants. *See under* Mental development
Intestines, 60–61
Iris, 138

Kidneys, 61

Leafy structures, 99–100
Learned behavior, 83–84
Lens, 139
Light, 132–135
Lightning, 126, 145–147
Liquids, 121–123

Machines, 153
Mammals, 96–97
Mathematics, 10–11
Measurement, 11, 40
Meat-eaters, 85
Melanin, 47
Mental development
  of infants
    newborn, 14–16
    at 4 months, 16–17
    at 7 months, 17–19
    at 10 months, 19–21
    at 1 year, 24–25
    at 18 months, 25–26
  of early schoolers
    at 5 years, 37–40
    at 6 years, 41–43
    at 7 years, 43–44
  of preschoolers
    at 2 years, 30–32
    at 3 years, 32–34
    at 4 years, 34–37
Microbiology, 109–115
Mimicry, 20
Minerals, 118–119
Mold, 109–112
Molecular theory, 148–149
Muscles, 49–52

Negative behavior, 32
Nerves, 63–64
Nose, 45, 57–58
Number relations, 43
Nutrition, 115–119
Nuts, 100–103

Observation skills, 29, 31, 36, 40
Object permanence, 20
Organisms, disease-producing, 113–114

Parasites, 89–90
Pecans, 101–102
Personality, 32
Photosynthesis, 98–99
Physics, 126–156
Plants
  dangerous, 108
  flowering, 107
  food from, 87
  leafy, 99–100
  photosynthesis of, 98–99
Poisons, 22
Preschoolers. *See under* Mental development
Prey, 86
Proteins, 116–117
Protons, 151
Pulse, 54

Rainbows, 135–137
Reasoning powers, 31, 41–43, 44
Refraction, 135
Reproduction, 65–66
Retina, 139
Rods and cones, 139–140
Roots, 103–104

Safety, 21–24
Scales, 74
Seasons, 94–97
Seeds, 104–107

Sex differences, 39–40, 65–66
Simple machines, 153
Skills
   balancing, 32
   classification, 20, 28, 40, 43
   communication, 40
   comparison, 34, 36
   measurement, 11, 40
   observation, 29, 31, 36, 40
   reasoning, 31, 41–43, 44
   See also Vocabulary
Skin, 47–49. *See also* Body coverings for
Smoking, 126
Social behavior, 21
Solids, 121–123
Sounds, 19, 127–129
Space, 153–156
Speech. *See* Vocabulary
Spring, 95
Stars, 156
Static electricity, 151
Substance misuse, 124–126
Summer, 95
Sunlight, 98–99

Teeth, 59, 61–63
Temperature, body, 114–115
Thunder, 145–147

Urination, 61

Vegetables, 100–103, 115–119
Veins, 53–54
Verbal ability, 19
Viruses, 113
Vitamins, 117–118
Vocabulary, 7–8
   at 10 months, 20–21
   at 2 years, 30–31
   at 3 years, 33
   at 5 years, 38

Warm-blooded animals, 96–97
Waste disposal, 119–121
Water supply, 119–121
Weather, 140–141
White blood cells, 115
Wind, 141–142
Winter, 95–96

Zoology, 71–97. *See also* Animals

Notes

Notes

Notes

Notes

Notes